The New Soundtrack

Volume 3   Issue 1   2013

Edited by

Stephen Deutsch, Larry Sider and Dominic Power

Edinburgh University Press

Subscription rates for 2013

Two issues per year, published in March and September

		Tier	UK	RoW	N. America
Institutions	Print & online	1	£100.50	£105.05	$188.00
		2	£126.00	£130.55	$231.00
		3	£157.50	£162.05	$285.00
		4	£189.00	£193.55	$338.00
		5	£214.00	£218.55	$381.00
	Online	1	£85.00	£85.00	$145.00
		2	£107.00	£107.00	$182.00
		3	£133.50	£133.50	$227.00
		4	£160.50	£160.50	$273.00
		5	£181.50	£181.50	$309.00
Individuals	Print		£34.00	£37.00	$67.00
	Online		£34.00	£34.00	$61.00
	Print & online		£42.00	£45.00	$82.00
	Back issues/ single copies		£18.50	£20.00	$36.50

How to order

Subscriptions can be accepted for complete volumes only. Print prices include packing and airmail for subscribers in North America and surface postage for subscribers in the Rest of the World.

All orders must be accompanied by the correct payment. You can pay by cheque in Pound, Sterling or US Dollars, bank transfer, Direct Debit or Credit/Debit Card. The individual rate applies only when a subscription is paid for with a personal cheque, credit card or bank transfer.

To order using the online subscription form, please visit www.euppublishing.com/page/sound/subscribe

Alternatively you may place your order by telephone on +44 (0)131 650 6207, fax on +44 (0)131 662 3286 or email to journals@eup.ed.ac.uk using your Visa or Mastercard credit card. Don't forget to include the expiry date of your card, the security number (three digits on the reverse of the card) and the address that the card is registered to.

Please make your cheque payable to Edinburgh University Press Ltd. Sterling cheques must be drawn on a U.K. bank account.

If you would like to pay by bank transfer or Direct Debit, contact us at journals@eup.ed.ac.uk and we will provide instructions.

Advertising

Advertisements are welcomed and rates are available on request, or by consulting our website at www.euppublishing.com. Advertisers should send their enquiries to the Journals Marketing Manager at the address above.

CONTENTS

EDITORIAL

This issue of *The New Soundtrack* features an interview with Peter Strickland, who discusses his recent film *Berberian Sound Studio* and how his working practices are influenced, among other things, by his career as a musician. Dominic Power's review of this much-praised film places the interview into context and should encourage any who have not seen the film to explore its remarkable facets.

Michel Chion's contribution centres on the complexities of synchronisation, reminding us of how synchronous sound and image is perhaps taken for granted. This strand is further developed by Philippe Ciompi, who asks us to listen to the remarkable sounds of everyday life, perhaps with our eyes closed. The desynchronisation thereby produced allows us more fully to appreciate the aural complexity of the world we inhabit.

Johan-Magnus Elvemo's article on multi-channel surround sound discusses the issue of 'immersion' into the filmic world, and explores the complexity of how the brain makes sense of a filmic world superimposed upon the physical space in which we experience film.

The sound world of the much watched and acclaimed American TV series *The Wire* is explored by Robert Walker. In his article 'Don't Pump up the Emotion: The creation and authorship of a sound world in *The Wire*' he explains the central role of music and sound effects to the notions of realism in the show.

Philippa Lovatt introduces many of us to the filmic world of Apichatpong Weerasethakul, and how through the use of image and sound, the Thai director expresses the 'process of remembering', through an amplification of the sonic elements so that they become almost 'denaturalized'.

The New Soundtrack 3.1 (2013): v–vi
DOI: 10.3366/sound.2013.0029
© Edinburgh University Press
www.euppublishing.com/SOUND

Finally, we offer an account of the School of Sound Summer Workshop, held with the ifs internationale filmschule köln, in August 2012. The purpose of the workshop was to develop more integrative approaches into post-production by encouraging editors, sound designers and composers to work together through all stages of a post-production project.

We hope that you find this issue stimulating. We also would like again to encourage readers to submit articles for future issues of this journal. In particular we are interested in the role of sound in documentary film, but would also welcome articles on any aspect of the integration of sounds with moving images.

PETER STRICKLAND

Sound at the centre: An interview with Peter Strickland

ABSTRACT

This article discusses the film Berberian Sound Studio (2012) in the context of the career and influences of its director, Peter Strickland. Decisions to focus the audience's attention on sound and the effect of using the audio track to visualise images is one of the most striking aspects of the film.

KEYWORDS

Italian horror film
1970's sound film
Berberian Sound Studio
Katalin Varga
The sound of cooking
musique concrète

Peter Strickland has combined his career as a film director with an interest in the possibilities of sound design and music. He is one of the core members of The Sonic Catering Band.[1] Born in Reading, he is Greek on his mother's side and British on his father's. He traces his fascination with cinema and sound to a visit to the Scala cinema at the age of sixteen to see David Lynch's *Eraserhead* (Lynch 1977). He started to make short films in 1990 and after a period of disillusionment he returned to filmmaking after receiving a legacy. In an interview with Ronald Bergan he described the dilemma as: 'I asked myself, "Should I buy myself a one-bedroom flat in Bracknell or should I make a revenge film in Transylvania?" I think the main thing that kept me going was knowing that if I bought a flat, I would always wonder, "What if?" Even if I failed, I would know I tried my

1. The Sonic Catering Band's core members are Colin Fletcher, Tim Kirby and Peter Strickland.

The New Soundtrack 3.1 (2013): 1–12
DOI: 10.3366/sound.2013.0030
© Edinburgh University Press
www.euppublishing.com/SOUND

2. *Bubblegum* 1996, directed
by Peter Strickland was a
fifteen minute short fea-
turing Nick Zedd, leading
figure in the mid-eighties
New York movement
known as 'Cinema of
Transgression' and for-
mer Andy Warhol super-
star, Holly Woodlawn.

very best.' After an initial struggle, his first feature, *Katalin Varga* (2009), was completed and released to critical acclaim. An essay analysing the film's soundtrack: *Soundscapes of Trauma and the Silence of Revenge in Peter Strickland's Katalin Varga* by Danijela Kulezic-Wilson appeared in this journal. His second film *Berberian Sound Studio* was produced by Keith Griffiths and has seen Strickland recognized as a unique and original filmmaker. Philip French in *The Observer* described it as 'One of the most remarkable British movies of the past couple of years', hailing Strickland as a key British film-maker of his generation. Peter Bradshaw's review in *The Guardian* described the film as 'utterly distinctive and all but unclassifiable, a *musique concrète* nightmare, a psycho-metaphysical implosion of anxiety, with strange-tasting traces of black comedy and movie-buff riffs.' Despite his British upbringing, Strickland is a very European filmmaker, exploring environments such as the Transylvanian badlands or the enclosed world of an Italian film studio where a violent horror film is in post-production. He currently lives in Budapest from where he did this interview with *The New Soundtrack* via Skype. Strickland proves to be an original interviewee, his conversation looping through a variety of topics including the propensity for coincidence in Dorking, the occult nature of vintage tape boxes, the glamour of past technology and Ben Burtt's sound design for *Star Wars* (Lucas 1977). The conversation's cast list includes Luciano Berio, Cathy Berberian, John Cage, Ennio Morricone, Bruno Maderna, Alvin Lucier, Joe Meek, Graham Bond and Alan Splet. For Strickland, sound is the key element of his filmmaking. 'There's this fear of silence,' he says. 'People equate silence with a lack of sound design, but that's not true. It's always the case of being sensitive to the material you are working with and what the film needs.'

The New Soundtrack: To go back a little, you have had a parallel career in music with your band The Sonic Catering Band. Which came first, film or music?

Peter Strickland: I made super 8 films from 1990 onwards; it culminated in a short film on 16mm that went round the festivals.[2] That was in 1996. I was kind of burnt out by that stage. I had been interested in music, though I felt that musically I wasn't very talented. But film just wasn't working out for me – I became so disenchanted with the cost and the hassle, but I still wanted to keep working and use some of those ideas in music. So we formed The Sonic Catering Band. For the next six years I didn't make anything in terms of film. Film and music were always together, but it was a case of practicality and what was available.

TNS: Sound design seems to be a recurrent preoccupation in all areas of your work?

PS: I'd always been fascinated by sound. Seeing *Eraserhead*, back in 1990, I got into music that way through Alan Splet's sound design. I'd grown up with mainstream films where sound did not necessarily convey a state of mind in the way Splet was doing. There's now this cliché of Lynchian sound, that industrial roar, and it's been appropriated quite badly. But when you hear that soundtrack there are so many layers, it's a world you can enter again and again. It's incredibly mysterious. Also I listened to bands like Stereolab[3] – their song titles, *Revox*, *Jenny Ondioline*, *Lo Boob Oscillator* really brought these machines to the forefront. I was also into the German bands like Can, Faust, Neu and Kraftwerk. Through Faust you get into *musique concrète*.

TNS: What was the concept behind The Sonic Catering Band?

PS: The idea was only to cook and make sound from the cooking. It was very important to have this documentary aspect, so we're not performing with the cooking. We are just making a meal and trying to ignore the fact that the mic is there. You do have to play it up a little bit, there has to be some kind of performative aspect. I wanted to make accessible music, not with the music itself but with the concept. Everybody gets the idea of cooking; it's like an entry point into what is difficult music. We try and keep the music as uncompromising as possible, but still have this simple notion of making the ordinary extraordinary.

TNS: So improvisation was an important part of this process?

PS: We liked the idea of a score – the recipe is our score. And there's the analogy of cooking and sound making. You're taking raw sounds and cooking them, and layering them and mixing them. When we started off we were too inexperienced and too excited by all the machinery. In hindsight we were only showcasing the machines not the actual sounds – and we were just drowning everything in reverb and so on. When I heard Alvin Lucier,[4] he got me really excited about how you don't need effects to make something work. *I Am Sitting in a Room* was such an epiphany for me, that he was making something so powerful just by rerecording and rerecording. Robert Ashley,[5] as well, just using editing and volume. That's when I felt confident that we could still use effects but we were more into just letting the sounds be what they are.

TNS: Did this process feed back into the way you used sound when you returned to filmmaking?

PS: My favourite recordings are where we ignored the concept and became more romantic about it, and concentrated on these lush soundscapes. We developed a certain way of working and people in our circle didn't always like what we did. But almost subconsciously we used that way of working in *Katalin Varga*. It was a pleasant surprise when people responded to it and this way of working was recognised. I didn't have any formal training, with the band it was just like a habit, to do sounds in this way.

3. Stereolab were one of the first "post-rock" bands of the 90s. They made extensive use of vintage electronic keyboards.

4. Alvin Lucier (1931-) is an American composer and a pioneer of experimental music and sound installations. His work has included the notation of performers' physical gestures, the use of brain waves in live performance, the generation of visual imagery by sound in vibrating media, and the evocation of room acoustics for musical purposes. *I Am Sitting in a Room* (1969) features a text read by Lucier, which is subjected to extensive rerecording.

5. Robert Ashley is a contemporary American composer, best known for his operas and other theatrical works, many of which incorporate electronics and extended techniques. Ashley was also a major pioneer of audio synthesis.

6. Vernon Elliott (1912–2008) was a bassoonist and composer. He was a founder member of Philharmonia Orchestra, a member of the Royal Opera House and part of Benjamin Britten's English Opera Group. In 1959 he began composing for the animator Oliver Postgate, providing music for much loved series for children, such as *Ivor the Engine*, *Noggin the Nog*, *Pogles' Wood*, *Pingwings* and *The Clangers*.

7. Basil Kirchin (1927–2005) was a British drummer and composer. Starting as a drummer in his father's big band, he moved on to scoring films and manipulating recorded sounds. He has been called 'the father of ambient music.'

8. Desmond Leslie (1921–2001) Anglo-Irish aristocrat, former spitfire pilot and author of books on UFOs. Desmond Leslie designed one of the first multi-track mixing desks during the fifties and began to experiment with *musique concrète*. His experiments in his home studio led to acetate called *Music of the Future* released in 1960. Leslie's experimental music appeared in some of the early episodes of *Dr. Who*. *Music of the Future* was re-released by Trunk Records in 2005. From 1945–1969) Leslie was married to the former wartime "black operations" broadcaster, actress and cabaret singer Agnes Bernelle.

9. Trevor Wishart (1946-) is a British composer who has been active in electroacoustic music since the seventies. He has experimented with tape manipulation and computer pieces. He works with the human voice, in particular with the transformation of it and the interpolation by technological means between human voice and natural sounds. Wishart has written two books *On Sonic Art* and *Audible Design*. He is currently

TNS: There's a contrast between the enclosed, claustrophobic world of *Berberian Sound Studio* – the tight shots, lots of close-ups, not much sense of the geography and the panoramic world of *Katalin Varga*. Was that intentional?

PS: Not necessarily. I had two or three projects ready to go after *Katalin Varga* and it was a question of which one it should be. I remember Keith Griffiths saying you should do something as different as possible otherwise you will get lumped in with a *Katalin Varga* type film all the time. It wasn't a conscious reaction against the first film, it was more the case of what was the story that fascinated me and what did it need. I see some similarities between the two films in how the atmosphere evolves. They both celebrate the spaces they're in; one celebrates the open space, one the inside space. Superficially there's not much difference apart from the suitcases.

TNS: In *Berberian Sound Studio* your protagonist is a sound technician. You are one of the few directors to incorporate sound into the drama this way. Coppola's *The Conversation* and Brian De Palma's *Blow Out* are the only other two films that really explore this idea.

PS: Gilderoy comes from what I would call that garden shed mentality of the semi-professional. Certain people came to mind in the conception of Gilderoy: Vernon Elliott,[6] who worked with Oliver Postgate. Basil Kirchin,[7] Desmond Leslie,[8] Trevor Wishart.[9] The idea of working with a Revox and razor in your garden shed is very attractive.

TNS: The title of the film is a tribute to Cathy Berberian?

PS: Definitely. It started with a friend called Michael Prime, who played me the album with Cage and Berio – it was the Berio/Berberian composition I really keyed into, *Visage*.[10] I never studied music formally, so I can only go on my emotional reaction. It was incredibly haunting and visceral. If it had been in a horror film or a thriller it would make perfect sense. It led into the idea of context and association. Take the way Penderecki's music was used in *The Shining*; on record the majority of people don't get that music. Just the idea of horror as a genre that really warms people up to ideas they wouldn't normally accept in a recorded medium. Conversely, if you think of Wayne Bell in Tobe Hooper's soundtrack for *Texas Chainsaw Massacre*;[11] if you played that in a concert hall, it's like Pierre Schaeffer. It's a form of lurid *musique concrète* but I guess people don't see it that way because it's associated with this hillbilly horror film. So Berio, Berberian and *Visage* were one of the starting points; just the idea of context and association. That also leads into the idea of innocent sounds, like vegetables been stabbed, which you normally do when you're cooking. But you put that exact same sound into a horror film it completely changes.

TNS: Cathy Berberian had a huge range, and the notion of the female voice is very important to *Berberian Sound Studio* because you focus in so much on the scream as a way of determining the actor's character and the character of the producer and the director.

PS: There was a fairly explicit tribute to Cathy Berberian with the witch coming from the dead. That was Katalin Ladik.[12] Maybe she wouldn't appreciate this but she was often dubbed the Yugoslav version of Cathy Berberian in the seventies. It was obviously very different from *Visage*. But I think people who know Berberian's work will see the connection between that sequence and *Visage*. I agree with what you said about the screaming and how people react to it. *Berberian* is quite an abstract film also it's quite visceral in the realistic sense in that if you keep screaming and screaming there's a physical consequence to that. Your voice is going to go. You don't normally think of that when you see a horror film. I wanted to show that.

TNS: There's a lot of recording kit from the period on display in *Berberian Sound Studio*. It makes the past recording technology very seductive.

PS: I couldn't have set *Berberian Sound Studio* in the present day, with plug-ins and laptops and Pro Tools and so on – it wouldn't be the same. What I love about that period, which only seems like yesterday, is that it's incredibly visual. You could happily watch people in the studio as an outsider. I wouldn't want to watch people working on a sound mix now. Peter Howell[13] helped me with some of the research and I remember him talking about the performative aspect of working with tape loops, cutting and splicing and just physically trying to reach things – standing up ladders and so on.

TNS: The studio was a visual environment back then. Now it can all be done on a laptop.

PS: I remember someone commenting, 'I don't know what these machines do, therefore it doesn't mean much to me.' Well, I don't understand what most of them do but find it even more powerful when you don't know. There's this alchemy there – they are this medium between the human being and the sounds that comes out. The notion of *musique concrète*, of taking everyday sounds and turning the ordinary into the extraordinary via this medium which is the machine, I find really fascinating. It also extends to the tape boxes back then. I remember talking to Julian House,[14] who did a lot of the designs. He has a great fondness for that period, and he was saying with a lot of the logos back then – if you were up too late at night doing tape loops, if your vision is slightly askew, they can somehow conjure up this pagan imagery, like sigils. It makes sense that people like Graham Bond[15] or Joe Meek[16] are associated with the occult.

TNS: How did you locate this equipment?

composer in residence at Durham University.

10. The American singer and composer Cathy Berberian (1925–1983) and the Italian composer Luciano Berio (1925–2007) were married from 1950–1964. *Visage* (1961) cut up and rearranged a recording of Cathy Berberian's voice.

11. Wayne Bell and Tobe Hooper are co-credited with the soundtrack for *The Texas Chainsaw Massacre* (1974).

12. Katalin Ladik is a Hungarian poet, performance artist and actress. She was born in Novi Sad, Yugoslavia, and emigrated to Hungary in 1992. She has created sound poems and visual poems and writes and performs experimental music and audio plays. Her work includes collages, photography, records, performances and happenings in both urban and natural environments.

13. Peter Howell (1946-) Before joining the BBC Radiophonic Workshop, Peter Howell performed with the folk group Agincourt. He began composing for *Dr. Who* in 1975 and in 1980 he updated Ron Grainer's original Dr. Who theme employing analogue synthesisers.

14. Julian House is a London-based designer who has designed album covers for numerous artists including Stereolab, Oasis, Prodigy, Broadcast and Razorlight. House describes his influences as coming from "the 'pulp' end of things which still informs me today, as much as classic design … Strange old vinyl LPs, paperback books, found ephemera …"

15. Graham Bond (1937–1974) was one of the godfathers of the British

Rhythm and Blues revival of the early sixties. His band The Graham Bond Organisation featured future stars John McLauchlin, Ginger Baker and Jack Bruce. Bond's later years were plagued by financial chaos, debt and an obsession with ritual magic. He developed a belief that he was the illegitimate son of Alesteir Crowley. He died by falling under the wheels of a train in Finsbury Park at the age of thirty-six.

16. Joe Meek (1929–1967) was a record producer and songwriter who pioneered the use of electronics. He wrote and produced *Telstar* by The Tornadoes. Meek's paranoia was exacerbated by his fears arising from his homosexuality (which was still illegal) and a growing obsession with the occult. He killed his landlady and then himself in his flat on the Holloway Road.

17. Studio di Fonologia Musicale was founded by Luciano Berio and Bruno Maderna at Milan Radio in 1954. It became a crucible for electronic music in Europe.

18. Bruno Maderna taught composition at the Venice Conservatory from 1947 to 1950. In 1950 he made his formal conducting debut in Munich. He subsequently became a great champion of the avant-garde. With Luciano Berio he helped to form the Studio di Fonologia in Milan in 1954. Also with Berio he was conductor of the RAI's Incontri Musicali from 1956 to 1960. He was chief conductor of the RAI in Milan from 1971. He composed the music for several Italian films beween 1946 and 1968.

19. Luigi Nono (1924–1990) Italian composer, who was one of the most prolific and influential

PS: It was very frustrating for us. I remember Daisy Popham, who did the props. She had a nightmare of a time trying to source that equipment because most of it disappeared and the rest is spread among collectors across the world. She did a great job but she really had her work cut out.

TNS: Did you see those machines as the mediation between the character and the world he arrives in? As if the actual personality of what he was doing was mediated by machines.

PS: Kind of – though I don't know if that's in hindsight, when we were editing it, or before. They definitely had a very magnetic power. I remember watching the Studio di Fonologia[17] footage – just on Youtube – with Bruno Maderna,[18] Luigi Nono[19] and Berio on it, It had this very spooky magnetic pull that draws you in. A lot of it was a desire to celebrate that. For dramatic effect, when we shot the machines, we didn't always use those exact sounds. In the two oscillator shots we're not using the actual sound, we're using tape delay on one and basic reverb on the other. So there's a lack of authenticity there.

TNS: There's a theatrical notion of the equipment itself.

PS: Definitely. We really played that up. Most people think that's what an oscillator sounds like, which it isn't. There's definitely a theatrical element in the tape loop – that goes into the whole notion of the film – the loop is central to the whole idea of the film. In hindsight I wish we'd done more with the blade, the cutting, because that feeds into the idea of the knife. We only scratched the surface and I hope other filmmakers come along and go further. It's not just analogue fetishism.

TNS: The iconography of creating sounds seems to work on several levels in the film.

PS: I'd seen that footage from Berio's studio and looked at album sleeve photographs. The Cramps label was putting out albums by John Cage and Walter Marchetti, David Tudor and so on – they had this great one, Morricone's band, Gruppo di Improvvisazione di Nuova Consonanza – *Musica Su Schemi*.[20] It's a great photograph of Morricone in this big studio, all these guys with their polo necks and their gongs and bowing instruments and so on. There's a chess board in the foreground with an orange telephone. And you start making up stories, what happened in their lunch break. You can just see a whole narrative. And of course the Radiophonic Workshop, which was a very different atmosphere but there was something very sinister in those machines. I was from that whole generation that grew up with *Doctor Who*, but it fed into that whole thing, though it's a very English preoccupation.

TNS: And what about the influence of *gialli* and that tradition of Italian horror movies of the 70s? Although we don't see any actual footage of

Giancarlo Santini's *The Equestrian Vortex*, apart from the lovingly authentic title sequence, it really conjures up the films of the period.

PS: Some of the films are incredible. When Dario Argento moved into things like *Suspira*,[21] it became very psychedelic which I really loved. I'm not a great horror fan as such but what I loved about that period in Italy was the incredibly heightened and cinematic style. They really celebrated all the elements; lighting, set design, sound, music. There was also the originality of the soundtracks; Morricone goes without saying – Ennio Morricone and Bruno Maderna really linked it to the avant-garde. I don't know this for a fact but it felt as if these guys were doing their very advanced *musique concrète*, the tape compositions and the electro acoustic stuff and on the side they were doing soundtracks for money. And maybe because academia wasn't paying attention to their moonlighting work they felt they could free themselves up. And they made stuff just as advanced, just as beautiful. Take *The Bird With the Crystal Plumage*[22] soundtrack, there are so many elements – *musique concrète*, dissonance free jazz – it's a very atonal soundtrack but very beautiful. Nothing jars, it's very cohesive. It's just truly wonderful. It's quite rare to find that now.

TNS: How fully had you conceived the film within a film?

PS: Up to a point, because it intersects with Gilderoy's life and the nature recordings and the shed in Dorking, I felt that we didn't know whether what we are seeing is *The Equestrian Vortex* or Gilderoy's mind, and I wanted to keep that ambiguity. So I kept the story up to a certain point, where the witches are still wreaking revenge. It's a simple rip-off of Argento's *Suspira* or Mario Bava's *Black Sunday*.[23] Some teenage girls are sent by their parents to a riding academy during the summer, it's run by this very stern mistress, and there's this equestrian library where they find this treatise on witchcraft. They find out that witches were interrogated in that building and buried in the poultry tunnel underneath, and they unwittingly unleash a spell which brings the witches back to life and then the film itself goes through these flashbacks of what happened hundreds of years ago. It's kind of interesting, this poultry tunnel thing was originally a slight tribute to *Death Laid an Egg*[24] which Bruno Maderna did the soundtrack to. When I was going to Dorking to read up about it, it turns out the chicken is the symbol of Dorking and the Dorking chicken comes from Milan, so there are all these coincidences. *The Equestrian Vortex* goes to a certain point and then I stopped writing it because to finish it would kill it off. It would be too concrete and we'd write off the whole idea of Gilderoy's home life. We did shoot some scenes, which I felt were too on the nose. There was a scene where Gilderoy's mother is involved in the witchcraft, but I'm kind of glad we cut it out.

TNS: So there won't be a director's cut with that in it?

PS: This is the director's cut.

composers of avant-garde music in the twentieth century. He is credited with coining the term "Darmstadt" to sum up the music created by Nono and his contemporaries, Bruno Maderna, and Karlheinz Stockhausen.

20. Gruppo di Improvvisazione di Nuova Consonanza was formed in Rome in 1964 by Franco Evangelisti. Its members included Franco Evangelisti (piano), Ennio Morricone (trumpet), Mario Bertocini (percussion and piano), Roland Kayn (Hammond organ, vibraphone, marimba), Frederic Rzewski (piano), John Heineman (trombone), Egisto Macchi (percussion, celesta). The group experimented with the development of new music techniques through improvisation. It disbanded in 1980 after the death of Franco Evangelisti.

21. *Suspira* (1977) is one of the most successful films by Dario Argento. Its plot concerns a ballet student who discovers that the dance academy where she studies is controlled by a group of witches. The music is by the Italian progressive rock band Goblin.

22. *The Bird with the Crystal Plumage* (*L'uccello dalle piume di cristallo*) (1970) is the feature debut of Dario Argento and is considered to be one of the most stylish and influential of the Italian *giallo* film genre. Italian/American actor Tony Musante plays an American whose life is threatened when he witnesses the murder of a woman in an art gallery.

23. *Black Sunday* (Italian: *La maschera del demonio* also known as The Mask of Satan) (1960) is considered a forerunner to

the Italian Giallo films of the seventies. It was directed by Mario Bava and features British actress Barbara Steele as a medieval witch who is executed but comes back to life to exact revenge.

24. *Death Laid an Egg (Italian: La morte ha fatto l'uovo* (1968) is a baroque *giallo* set on a factory poultry farm. It was directed by Guilio Questi and starred Ewa Aulin, Gina Lolliabrigida, Jean-Louis Trintignant.

25. Broadcast are an electronic band that provided the soundtrack to *Berberian Sound Studio.* Its core members were James Cargill and the late Trish Keenan.

26. Andrew Liles is a sound artist and multi-instrumentalist. He has a vast output of recordings that he has released since the mid-1980s, covering a variety of styles as experimental music, dark ambient music, progressive rock. The artists he has worked with include Steven Stapleton, the Hafler Trio, Faust, Nurse With Wound, Daniel Menche, Band of Pain, Paul Bradley, Aaron Moore, Nigel Ayers, Frans De Waard, Freek Kinkelaar, Danielle Dax, Rose McDowall and Ernesto Tomasini.

27. Joakim Sundström is a sound designer, sound editor and musician. He was the supervising sound editor on *Berberian Sound Studio.* He has worked on numerous films including works by the Brothers Quay, Stephen Frears, Patrick Keiller, Ben Hopkins, Kevin MacDonald and Michael Winterbottom.

28. Chris Dickens was the editor on Berberian Sound Studio. His credits include *Shaun of the Dead* (2004); *Hot Fuzz* (2007), *Slumdog Millionaire* (2008) (for which he

TNS: You imply that you perceive the music track to some extent as separate from the normal sound design. In *Berberian Sound Studio* those elements seem completely fused. Was the post-production just a unity or were these things separate elements?

PS: It was pretty unified. The one rule I had was to make a diegetic soundtrack. When we were editing we knew it was very abstract, so we needed this sonic anchor to the film which is very realistic – even though we mixed this in a very non-diegetic way. So when the sound started quite early on, both Broadcast[25] and various friends handed in demos, just recording their kids screaming, their wives screaming, whatever. I recorded a couple of friends screaming as well and I sent those off to a guy called Andrew Liles[26] and that's when things really came alive. He sent back two CDs crammed full of treated screams, a mixture of very aggressive analogue distortion to very beautiful interweaved and quite abstract and atmospheric sounds. I played it on set and it was great to have it in the edit. It really informed how it went. It went back and forth; James Cargill from Broadcast would send mp3s over in emails and they would dictate how the edit would go and, vice versa, we'd send a cut back to James, so there was a constant to and fro. Other treated sounds came in from Steven Stapleton, Tim Kirby, Jonathan Coleclough, Colin Potter, Colin Fletcher, The Bohman Brothers and Clive Graham. In terms of the fusion of music and sound – it wasn't conscious as such, it's just the way it ended up. Because it was diegetic it had a kind of flatness to it. Joakim Sundström[27] suggested we do a basic stereo mix for Santini's film. So we tried to keep within that to some degree. The main thing all of us agreed on was perspective; that you take one sentence and obviously you're cutting between a mouth in the mixing room and that voice coming through a tannoy or headphones. Many things were discovered in the edit. One thing I was very certain of was that we would need many pickup shots which could really help us with the sound mix. So we shot, at random, a lot of shots of Toby's hands on the mixing desk, oscillators and so on. We almost used it like pepper when you're cooking. There was no plan; Chris Dickens,[28] the editor is really keen on using sound as much as possible. If we wanted a hard cut for dramatic effect, we would go through the pickup list and say 'right, he's punching that sound out, he's muting this, let's put that in there'. It was very liberating. I remember with *Katalin Varga* we had to think about how do we gently fade the sound out and how do we fade out the mood music and normally to give a hard cut you need a very dramatic reason for it. With *Berberian*, it's what the characters do, so we had a lot of freedom – it was very exciting for us to work that way. The other thing was subjectivity, and how far we achieved it. For instance, when you hear dialogue coming through the actress's headphones, the frequency's very narrow. When it comes through Gilderoy, he has his headphones on at full volume and it's there for the audience. We're obviously paying attention to his point of view. So we're cheating it sometimes.

TNS: Was there a soundtrack to which Chris Dickens was editing, or was it being constructed during the edit or post edit or all of those three?

PS: All of those things. There were very few tracks when we started. We had the abstract screams which were very useful. It was a very difficult time for James because his partner died at that point – it's a miracle that he actually kept producing music. Music came a few weeks after the assembly edit. At that time we knew that we had to use temp tracks. We used mainly Morricone tracks – it's a risk because one becomes attached to it, and financiers become attached to it. Especially with that music because it's so lush and so beautiful. We were still mixing things after the edit. Chris had a fairly rough track laid down. When that came to Joakim we were still working on it. There was a final push with Christer Melén,[29] we did a lot of stuff in his flat. We then took it to the final mixing stage with Doug Cooper and Markus Moll. Everything came alive in a very satisfying way when Doug and Markus were at the controls, and Linda Forsen's dialogue editing helped enormously. It depends on different teams but I'm always putting things down at the last minute. I find it hard to come into a sound studio with a complete lockdown on the track laying.

TNS: Was the structure of the film itself, or sections of the film, determined by how you felt you would be dealing with the sound?

PS: Definitely. Some of the vegetable shots, they were not in the script, and in the scene with the footsteps that was a slight tribute to Morricone's *Short Night of Glass Dolls*,[30] with a Watkins Copicat[31] – that was all done before hearing what James had done.

TNS: The enormous tape loop that goes through the studio, clearly you had to shoot it for that?

PS: That was in the script, but we didn't have a sound in mind at that point. So that went through two stages. First it was an organ piece, which was not looped, the organ was just on the mixing desk. The fire was looped – it was basically a witch being burnt at the stake and a fire was looped. A looping fire doesn't sound that interesting so we started multi-tracking these prayers and having them go loud again and again which has a kind of whirlpool effect to it.

TNS: I suppose that takes us to the last ten minutes of the film. Because here it's about the relationship between the sound of what is inside and what is outside the studio. Were you treating sound in a different way? Were you changing the rules of perspective, for example?

PS: No, I think we were pretty consistent. The only thing we did was use a slightly more jarring volume. All those sequences with projection at the beginning of the film, they were quite soft and then we started to do a Mid Sweep on it. For me the turning point is when Gilderoy refuses to squirt the water, when the horror becomes so extreme. We started using a bit of Mid Sweep. It just takes on a more rasping quality. It just becomes louder and louder, and the analogue distortion becomes more

received an Academy Award for Film Editing and a BAFTA Award for Best Editing) and *Les Miserables* (2012).

29. Christer Melén's work as a sound effects editor includes Gideon Koppel's *Sleep Furiously* (2008), Andrea Arnold's *Fishtank* (2009), Michael Winterbottom's *The Killer Inside Me* (2010), Stephen Frears' *Tamara Drewe* (2010) and *Seven Psychopaths* (2012) by Martin McDonagh.

30. *Short Night of Glass Dolls* (Italian: *La Corta notte delle bambole di vetro* (1971) is an Italian *giallo* directed by Aldo Lado, with music by Ennio Morricone. It starred Ingrid Thulin, Jean Sorel and Barbara Bach.

31. The Watkins Copicat is a Tape-Echo Unit designed by Charlie Watkins in 1958. It is the first repeat-echo machine manufactured as one compact unit. Its tape-echo sound had an enormous impact during the late 1950's and early 1960's, and was behind the guitar sound of the top British bands including The Shadows.

32. In Brian De Palma's *Blow Out* (1981) John Travolta plays a sound effects technician, who captures evidence of a planned political assassination, which deals with themes explored earlier by Michelangelo Antonioni in *Blow-Up* (1967) and by Francis Ford Coppola in *The Conversation* (1972).

33. Peter Cusack is an artist and musician who is a member of CRiSAP (Creative Research in Sound Arts Practice), and is a research staff member and founding member of the London College of Communication in the University of the Arts London. He was a founding member and director of the London Musicians' Collective. He was part of the *avant garde* quartet, "Alterations" from 1978 to 1986; with Steve Beresford, David Toop, and Terry Day. He is the creator of field and wildlife recording-based albums.

34. Christopher Richard Watson is a founding member of the group Cabaret Voltaire. As a wildlife sound recordist he has covered television documentaries and experimental musical collaborations.

unpleasant – though actually I find it quite pleasant. But otherwise, even though Toby's voice is dubbed, we still tried to keep to that perspective.

TNS: What about when he's in the studio, seeing the film of him waking up. Did you treat sound differently there?

PS: No, not really. Initially we tried to make it more eerie; we multi-tracked a lot of whispered prayers and some ambient sound. And we thought it was too much, it was counter-productive. We pulled everything back and let the visuals do the work.

TNS: Did you manage to get all you wanted within the constraints of the budget?

PS: When you do a film obviously there's a lot of money involved. So you do your trade-offs. There were some minor things I traded to get more important things the way they should be in the film. If I went back to it I'd have a bit more machinery in it, a bit more tinkering around with the physical aspect. I remember watching *Blow Out*.[32] I loved those scenes with Travolta, going back and rocking and rolling. That's just me. I love all that. Of course a lot of it wasn't available. With the time frame we had we had to get hold of some fill-in gear. Luckily we have some great machines in there. A lot of it is quite personal to me, like the Watkins Copicat. We've used it in the band since 1996. We used the very same model of the Copicat to make that feedback and those tape loops.

TNS: What was the shooting format?
We shot *Berberian* on the Alexa camera, except for a couple of sequences that were shot on 16mm. We created the sound on digital means, we weren't using analogue. I was never an analogue purist as such. But what I object to now is that digital is being forced upon us. People are not given a choice any more. With Digital projection I prefer the 2K hard drives. I remember seeing prints of *Katalin Varga* that varied so much, some of the sound was appalling on 35mm. So as a personal preference 2K is great, but I really object to the industry being cornered into using digital.

TNS: How did you set about capturing the sounds?

PS: Obviously when you do a film on a low budget we had a lot of time in the studio, but there was no money to record outside. Joakim has this plug-in which can replicate any kind of room tone or natural reverb outside. I'm not convinced by them, there's nothing like going out there and recording them yourself. When I listen to recordings by Peter Cusack[33] or Chris Watson[34] – you just cannot do those in the studio, you just have to go out there with a microphone. It worries me, because the technology is so good, the downside of that is that there is the risk of a kind of laziness in attitude that would potentially go all the way from sound people to producers. If you go back to the seventies, consider the amount of time that even mainstream people were given. Think of Ben Burtt[35] with *Star Wars*, using completely experimental methods for a mainstream film. I think Lucas gave him a year. It's great to read about what he was doing and how he experimented. Alan Splet as well. I remember having an argument with someone a while back about library sounds. His argument was, 'Who's going to find out? We've got this library of sound effects, no-one's going to know.' That's not the point. The point is that process of discovery. It goes back to sound and association. My most visceral experience of this was being punched in the face when I was at school and being surprised at how undramatic it sounded – I was so used to hearing punches on TV. In *Berberian* stock sounds had a place, we did use them, because of Santini's film; Santini would have used stock sound effects. In *Katalin Varga* it was essential not to use them. Even if people don't notice, you as a director, you notice, when they come off the shelf, they come with a baggage. Unless there's a dramatic reason for it, it's just not right. There's always the argument that you can get it online and hence it's easier and cheaper. On *Berberian* someone suggested that for the photograph of Gilderoy's shed, I should get it on line or go to a photo library. You don't need to go to Dorking to find a shed, etc. For me it was very important *to* go down to Dorking, to find someone with a shed and to know that it actually is below Box Hill. The beauty of doing it was I met this person, who asked what we were doing, and I told him it was for this film and it turned out he was a fan of Broadcast, who did our music. You get these lovely coincidences this way. We had a very generous amount of time to do mixing, but it worries me that sound is always last.

TNS: What projects have you been working on recently?

PS: We were in the Czech Republic last week for a festival for John Cage's centenary. Jozef Cseres,[36] who plays Massimo in *Berberian Sound Studio*, is a huge John Cage person – he met him back in 1990 just before he died. He always wanted us to do performances of his macrobiotic recipes. I didn't want to do it originally. It didn't feel right to be doing something on the coattails of someone else. However, this year felt like a good excuse to do it, because of the centenary. If ever it was acceptable, this was the year to do it. So we did these concerts, I haven't heard them but it's very different music when you play live with cooking.

TNS: Doing John Cage's recipes is a new way of doing covers.

PS: Yeah, they're cover versions, we did the Miso Soup, mushrooms of course, we did his zucchini recipes. It was a really good event. We're doing an album of it. Jozef has written a book about Cage and the recipes and we're going to compile some of the live recordings.

TNS: So you continue to combine film with music?

35. Ben Burtt (1948-) was the sound designer for the *Star Wars* and the *Indiana Jones* films. He is responsible for some of the unique sounds associated with *Star Wars* including the voice of R2D2, the sound of the light-saber and the heavy breathing of Darth Vadar. Among his other credits are *Invasion of the Body Snatchers* (1978) *E.T. The Extra-Terrestrial* (1982) and *Wall-E* (2008).

36. Jozef Cseres played Massimo in *Berberian Sound Studio*. He also writes on contemporary music issues and organised the 'Sound Off' festival in Nove Zamky. He's worked extensively the Fluxus artist, Ben Patterson as well as Otomo Yoshihide and Sachiko M.

PS: Film and music were always together. But with film I was more confident. With sound I could never sit in a studio by myself – I just know the basics. I remember when I first came into the studio I was terrified, because one of the band members couldn't get out of bed, and he was the one who knew what to do technically. It was a disaster, I just couldn't use anything. But then I slowly learnt. You only use 20% of the studio, it's like a computer. Once I got that in my head, I knew the basic tools. Still I would never trust myself to work on my own. So it's mainly ideas. I work with people I trust who are very good and sympathetic. There's Joakim; there's Gábor Erdélyi Jr., who worked on my last film and who works with Bela Tarr. There's Colin Fletcher and Tim Kirby from the band. They're co-pilots basically, they're so essential to the whole process.

TNS: You appear to have managed to keep a degree of independence.

PS: With *Berberian* it gave me some hope that one can continue working this way. Obviously I'd always like longer, just to have more time to go out there and physically record things. It's the joy of the process. Some people misunderstand directors; they think we like the red carpet. Some directors do but for a lot of directors their red carpet is to go out there in the middle of nowhere and record something. That is our red carpet, the joy of discovery. We wouldn't do it if we could just have an application that would make a film. The whole purpose is to have that fun making it.

DOMINIC POWER

Berberian Sound Studio

Another tramp told the story of Gilderoy, the Scottish robber. Gilderoy was the man who was condemned to be hanged, escaped, captured the judge who had sentenced him, and (splendid fellow!) hanged him. The tramps liked the story, of course, but the interesting thing was to see that they had got it all wrong. Their version was that Gilderoy escaped to America, whereas in reality he was recaptured and put to death.

George Orwell *Down and Out in London and Paris* (1933)

Peter Strickland's bewitching *Berberian Sound Studio* opens with one of horror fiction's classic tropes: a weary English traveller enters an unfamiliar environment, his host, a flamboyant foreigner, bids him enter with the greeting, 'Welcome to the world of sound.' Strickland may have replaced the badlands of Transylvania that formed the backdrop of his break-through debut, *Katalin Varga* (2009) with a more enclosed world for his second feature, but his opening nevertheless conjures up a more familiar Transylvanian association. In Bram Stoker's novel *Dracula* and in Todd Browning's 1931 film version, the English traveller Jonathan Harker is similarly summoned on a mysterious mission and, on arrival at Castle Dracula, is greeted by the Count with another invocation to sound, 'Listen to them – Children of the night. What music they make.'

The setting is an Italian sound studio during the mid-seventies, at the height of the *Giallo* boom – the exotic horror genre that flourished in Italy from the late sixties onwards, and whose most notable practitioners were the film directors Mario Bava and Dario Argento. Into this baffling and occasionally sinister environment comes Gilderoy, a shy, reclusive Englishman played by Toby Jones. Gilderoy is a sound engineer,

The New Soundtrack 3.1 (2013): 13–15
DOI: 10.3366/sound.2013.0031
© Edinburgh University Press
www.euppublishing.com/SOUND

summoned by the producer, Francesco Coraggio (Cosimo Fusco), to mix the sound on *The Equestrian Vortex*, a mix of sadism, sex and Satanism set in an all-girl riding academy. Questions hang over Gilderoy's presence in the Italian studio; why has this obscure semi-professional, whose previous experience has been limited to wildlife films and children's television, and whose studio is a garden shed in Dorking, been summoned by a cash-strapped Italian film company to oversee the post-production sound?

The studio is a world unto itself, where we never know whether it is day or night. The atmosphere becomes increasingly hostile, unmediated by the realities of the outside world. After an initial flourish of bonhomie, Coraggio becomes a tyrant, refusing to reimburse Gilderoy for his plane fare and subjecting him to a barrage of petty insults. The director, Giancarlo Santini (Antonio Mancino) is a narcissist, convinced of his own genius ('This is not a horror film. Is a Santini film') who preys on the female members of the cast. The rest of the crew are either indifferent or openly hostile. Soon the atmosphere is heavy with sadism, as if the inhabitants of the studio are becoming corrupted by living with repeated scenes of horror. Gilderoy is a herbivore in an enclosed world of carnivores, co-opted into destroying a vegetable matter to create the sounds of torture and death: the squelch of a dropped marrow stands in for a body hitting the ground, radishes separated from their stalks become hairs being pulled out during interrogation, and the sizzle of frying fat represents the sound of a hot poker against flesh.

Gilderoy appears to be protected from the air of corruption by a home life which he carries with him – his makeshift studio in the garden at Dorking and life with his mother in the English countryside. Letters arrive from his mother, full of loving details about the new nest of chiffchaffs, letters which he is never allowed enough time to answer. These letters take a darker turn as the studio itself, and Gilderoy's role within it, become more unreal.

Stickland's bilingual script plays on ambiguity and menace. The dialogue is littered with threatening non-sequiturs – when Gilderoy mildly mentions a technical problem when he first meets Santini, he is brutally rebuffed by Coraggio with the mystifying put-down, 'When you first meet someone for the first time, try and discuss only positive things. It is good manners.' Both producer and director insist on embracing Gilderoy on first meeting only to treat him with impatience and contempt moments later.

Berberian Sound Studio avoids easy categorisation. It is a horror film in its sense of brooding menace and invocation of unseen cruelty. It is also a film about the process of film, showing the mechanics of post-production while referencing a variety of other films. It also explores the nature of sound; except for a brilliantly realised credit sequence we never see any actual footage of *The Equestrian Vortex*. Though we can see the process of the sounds being created, they do not lose their power to suggest terrible things. The vocal effects, created by Katalin Ladik as the reincarnated witch and Jean-Michael van Schouwburg as a 'dangerously aroused goblin', transcend their purpose and attain an eerie sonic beauty. All the while the camera lingers over the period sound equipment, investing them with a sheen of glamour. Even the dubbing sheets look like powerful abstract art.

The film inhabits the territory between high and low art, where the avant garde and the world of exploitation collaborate both deliberately and accidentally. It shows the mechanics of creation and the effects of alienation and isolation. Its evocation of claustrophobia and unease calls to mind other movies that move in the same psychic territory: Roman Polanski's *The Tenant* (1976) for its charting of the psychic landscape of the outsider, or Coppola's *The Conversation* (1974) for the way it looks at the deceptiveness of sound, or David Lynch's *Lost Highway* (1997) or *Inland Empire* (2006) for its beautiful warping of reality. But *Berberian Sound Studio* transcends its influences to create a truly original work that lingers in the mind long after the film has ended.

The distinctive name of its central character is just one of the puzzles in *Berberian Sound Studio*. It appears most recently as belonging to the celebrity wizard Gilderoy Lockhart in Harry Potter. However, there is an older, more resonant Gilderoy – a legendary Scottish outlaw who featured in numerous ballads – a swashbuckling character who at first seems an unlikely antecedent for the timid protagonist of *Berberian Sound Studio*. However, in *Down and Out in London and Paris*, George Orwell points out that in the retelling of the story among British vagrants, Gilderoy's authenticity and his true nature and destiny are changed through the process of sound, recreating him as something else entirely.

Berberian Sound Studio

UK 2012 92 mins.
Director: Peter Strickland; script: Peter Strickland; Director of Photography: Nicholas D. Knowland; editor: Chris Dickens; supervising sound editor: Joakim Sundström; sound effects editor: Christer Melén; production design: Jennifer Kernke; art direction: Sarah Finlay; producer: Keith Griffith, Mary Burke.
With: Toby Jones, Cosimo Fusco, Antonio Mancino, Tonia Sotiropoulou, Susanna Cappellaro, Eugenia Caruso, Katalin Ladik, Jean-Michael van Schouwburg.

Dominic Power is an editor of this journal

MICHEL CHION

The 'clap'
Translated from the French by Don Siegel[1]

ABSTRACT

The synchronous recording of sound onto moving images has been accepted as a 'given' by most audiences and film-makers since reliable technology for its production appeared in the late 20s. However, the relationship between such image and sound synchrony is quite complex. Michel Chion discusses this relationship in the context of film's history and of his own dictum: 'There is no soundtrack'.

KEYWORDS

Synchronisation
Clapperboard
Fellini
Godard
Audio-visual
  relationship

Fellini's celebration of cinema, *Intervista* (1987), ends with a sharp 'clap' that shows a new film is about to begin shooting. While few people know the actual purpose of the clapperboard and its distinctive sound, it has nonetheless long symbolised the Seventh Art.

Before becoming a symbol, the clapperboard already had a very concrete purpose. The sharp 'clap' was recorded by the location recordist. By matching it with the visible action of the two pieces of wood being struck together (filmed by the camera), it could be used to synchronise picture and sound. In other words, the clapperboard alone represents the advent of the talkies since it brought together their two constituent components.

1. This article was first published as *Les liens du son*, in the monthly *Le Monde de la Musique*, February 1988, number 108. The original article in French is appended below.

The New Soundtrack 3.1 (2013): 17–25
DOI: 10.3366/sound.2013.0032
© Edinburgh University Press
www.euppublishing.com/SOUND

Today, this type of audio-visual synchronisation seems so self-evident that it is unworthy of the slightest cinematographic interest. In particular, this viewpoint would be shared by all those who underwent anti-synchronist brainwashing in the 1970s – an ideology that scorned any literal correspondence between sound and picture and, as in the films of Marguerite Duras, celebrated floating words and lost sounds. From this viewpoint, the clap was the very symbol of the humiliating slavery that had until then kept sound shackled by the image.

But synchronisation wasn't always seen in this way. For the first audiences to see talkies, at the end of the 1920s, it was quite the contrary: an absolutely marvelous invention. Ads boasted about the synchronisation of films just as the latest high-fidelity sound systems are plugged today. We have to assume that any previous talking picture experiments, like Gaumont's *phonoscènes*, shown in Paris from about 1906 to 1912, fell far short of the mark. Furthermore, just at the time talkies came along to disrupt things, a few inventors had become obsessed with finding ways of more reliably synchronising the silent film and the live music accompanying it. This magnificent obsession led to the invention of various mechanical and metronomic systems to guide better the conductor or pianist, so that the musical effects would come at the exact moment desired. The search for synchronised sound and picture – not yet automatic and obvious – in fact inspired a desire as passionate as the search for perfection by an ensemble rehearsing a piece of chamber music.

And yet, the fact that a sound can be heard in unison with an observable image is not really that obvious, even if we see it thousands of times a day. The sound naturally lags behind the picture, for two reasons. First, because it is often an effect, and not the cause of the image, and thus comes after, even if only by a millisecond. Secondly, because of the slowness of propagation: sound moves at 340 metres per second in the air, or a million times slower than the speed of light! The apparent pairing of sound and image is therefore only an approximation of our human perception, but this approximation is the very basis of our experience: even a child in the cradle is immediately sensitive to the synchronisation of oral and visual phenomena. In movies, a sound we match to a visual movement will be automatically associated with it (the principle of sound effects), even if the sound's content, color and tone are only roughly in synch, or even totally out of synch with the phenomenon illustrated.

This synchronisation of sound and image, symbolised by the clapper-board, was a wondrous phenomenon for the nascent talking pictures, a magical meeting of two principles that produced the spark of life. Perhaps this is best illustrated by some of the first cartoons from Walt Disney, those amazing *Silly Symphonies* from the early 1930s. These are musical shorts with wonderful miniature concerts given by mice, ducks and other animals, on extravagant instruments that could even take on a life of their own, and play themselves in a wild performance with absolute perfect synchronisation between the audio and the visual. This was only natural of course: each image was drawn to correspond to a pre-recorded sound, a technique that is commonly used in animated film.

It was a logical response for the new-born talking pictures to focus on the concert as a show. Nothing could be more apt to illustrate the law of synchronisation than instruments – a series of micro-actions, small visible shocks, each of which produces a distinct sound. So it's hardly surprising that the dawn of sound cartoons could find no better way of singing the praises of synchronisation than showing animated concerts. Take for example *The Birthday Party* (Gillett), from 1931, a wild celebration of Mickey Mouse's birthday that is also a synchronisation party. Mickey's animal friends have offered him an upright piano, which he immediately plays like a virtuoso, with his famous four-fingered, gloved hands. Then Minnie sits down at another piano, and our favorite mice play, back to back, a frantic duo that would make the Labègue sisters look rather staid in comparison. Other fantasies in the film include a xylophone that comes to life on its four legs, like a horse, and begins to play itself, with the different keys leaping up and falling back in just the order needed to produce a furious piece of music, synchronised note for note. The parallelism of the sound and image is here constructed lovingly, like the movement in a Swiss watch, in the ravishing new world of synchronisation.

Ironically, the dynamic symbol of the clap, that energetic union of sound and image, is offered to us by Fellini – the cinéaste who, more than any other in the history of movies, employed a floating poetry and very approximate connections between sound and image. Like any self-respecting Italian filmmaker he dubs all his films – there is no direct sound recording. But in dubbing, he purposefully leaves a margin, a slight undisciplined gap between the movement of the lips and the post-synchronised voices.

However, strictly speaking, this has nothing in common with the disembodied voices – an over-the-top voiceover – in the last films made by Marguerite Duras. In Fellini's films the voices in fact match the bodies, but not very smoothly. Only in *Casanova* (1976) did Fellini allow himself to push desynchronisation to its limits: the dubbed voice of the seducer (in the original Italian version, the actor Luigi Proietti) is offset by several seconds from the visible movement of the lips of the Canadian actor playing Casanova, Donald Sutherland. Even in this case, however, we're still in the world of synchronised films: the movie simply seems to be taking place in a different kind of environment, a dreamlike and murky atmosphere where sound is significantly ahead of or behind the image.

There are in fact two basic approaches to the relationship between sound and image, and here I am referring to directors who have totally revamped their art, from Welles to Tarkovsky, and from Bresson to Tati.

On the one hand, there are the 'dualists', who see sound and image as a couple comprising two separate, basic units, and therefore can only conceive of there being love or hate between them, absolute union or separation: Godard and Duras can be placed in this category, for instance, or, looking further back, Pagnol and Renoir. Then there are those who see sound and image as simply two poles that, by coming together, create something other than sound and image, something irreducible to one or the other, or to one and the other. They create an effect, an energy, a current, a feeling. Fellini is a leader of this camp, what I call the 'triangulists', as opposed to the 'dualists'.

The same year that the Fellini film came out, there was another tribute to cinema from a celebrated director: Godard's *Soigne ta Droite* (*Keep Your Right Up!*, 1987). The dualism of the auteur is already signaled not only by the title, but also by the poster, which shows two hands, one holding the other, which are immobilized by handcuffs. The same director who made films called *One Plus One* (1968), *Numéro Deux* (1975) and *France/tour/détour/deux/enfants* (1977–78), here once again addresses the question of the relationship between sound and image like a couple in conflict. Sound and image chase each other at high speed, each enclosed in its own being. Each brilliant effect in the movie – editing of the sound and picture, and the sound that is offset in relation to the image, behind or ahead – falls like a blow: it's like a point scored or lost, always provisionally, in a fight or in an embrace. The 'clap' in Godard's films (and there are a number, because he was one of the first directors to show films within films) can only be a bell announcing a new shoot, and not, like Fellini, a spark giving rise to a new vision of the world, whose constituent elements of sound and image are mixed, even transfigured.

2012 AFTERWORD: THE AUDIO-VISUAL RE-DIVIDED

The above text was written in 1988 for the monthly *Le Monde de la Musique*, at the same time as the release of *Intervista*, by Fellini, and *Soigne ta droite*, by Godard. I offered this piece to *The New Soundtrack* not only in an attempt to rehabilitate the poor step-child of film theory known as synchronisation, but also for its historic interest. At that time, things seemed simple: there appeared to be space for a dual division of cinema into sound and image. Of course, I had already sought to break up their false symmetry, by showing (in *La voix au cinéma*, 1982), that 'there is no sound track' and that it's more accurate to define cinema as a space for images on one hand (a space defined by a frame), and for sounds, on the other, sounds that don't have their own frame. But I still considered the sound/image division as pertinent.

As early as 1990, however, I believed that a third element came into play: quite simply the 'text', present in 99% of all films in a dual oral/written form, and not addressed by most cinematographic analyses. Gilles Deleuze was very successful in unifying cinema under the notion – vague in my view – of 'image', which enabled him to define sound as a 'component of the image' (*L'Image-temps*, chap. IX). For him, the word 'image' was like 'The Blob' in a science fiction film, swallowing up everything. It's not a bad idea to unify cinema via a key concept: in the writings of Eisenstein and other Russians the concept of 'montage' had played this role, seeking to go beyond the technical opposition between sound and image. But a downside emerged when we tried to apply this concept to everything. For example, using 'montage' to describe simultaneous relations (the 'vertical montage' that would be generated between a sound and an image) would be an incorrect use of the term.

I tend to go in the opposite direction, namely a 're-division'. In my article from 1990, *L'Audio-vision*, my analysis led me to the concept of an *audio-logo-vision*, because language cannot be associated solely with sound, if it involves voices, or with an image, if it involves written characters.

So what about a division into three: text, image and sound? No, because it would be arbitrary to study a text pronounced or read, independently of the rest. We can also divide the cinema into 'shown' and 'said'. 'Shown' refers to what is shown in concrete form as sounds and images. 'Said' is what is formulated and verbalized by characters, in voiceovers, in songs, etc. I make a distinction[2] between five main said/shown relations: *punctuation, deepening, contrast, counterpoint, contradiction*. The most often used, even today, is the first. It involves having the words pronounced by characters listened to with the help of actions and events that have no relation with the basic content of what is said: cigarettes smoked by the characters, the objects they handle, food and drink, certain games and sports – in films, these provide infinite opportunities to punctuate what are often abstract dialogues. Some directors have refused this process, and practice a 'non-punctuated' cinema, like Éric Rohmer, or Kubrick in several scenes from *2001*, whose characters do not act when they speak, and do not speak when they act.

Can we therefore reduce cinema to two components, the 'said' and the 'shown'? It's not that simple.

It's worth noting that sound and image share some aspects, including that of rhythm. There is not a special 'sound rhythm' or a special 'visual rhythm'. Even if the ear can pick up rhythms faster than the eye, the aspects of slowness/speed, regularity/irregularity, steadiness/acceleration/deceleration that may characterize a given rhythm are the same for the eye and the ear. Rhythm is a 'trans-sensorial' notion, i.e., one for which the conventional division into the five senses does not apply.

Is the text considered a non-divisible aspect? Of course not, because a read and heard text (sound film uses them both) is fundamentally different, and this difference is not comparable to the difference between sight and sound. Texts are read at a speed of our choosing (even though limited by the film), while hearing is subject to the pace of the words being spoken. Furthermore, reading brings us into contact with silent signs and symbols (periods, for instance, but also letters that may not be pronounced in certain languages, like the silent 's' in French plurals), which are nonetheless part of writing.[3]

Lastly, there is what I call the *audio-division*, by which I mean the division created 1) inside what we see on the screen, by what we hear from the loudspeakers, and 2) inside what we hear, by what we see on the screen. Depending on whether I can see the source of the sounds I hear, the sounds are divided into 'onscreen sound' or 'acousmatic', and take on a different meaning. Depending on whether I can hear the sounds corresponding to the actions, objects and characters that I see on the screen, these sounds seem different to me. The sounds re-divide the image in the film, the image re-divides the sounds in the film.

2. See *Film, A Sound Art* (Chion 2009).

3. My current research is focused on this subject, namely the role and position of written texts in cinema, thanks to a grant from the IKKM in Weimar.

The idea of these re-divisions, which came to me as I studied a number of films, is different from a subdivision that would continue the existing categories (sound/image), or that would multiply the defined distinctions in indefinitely divisible sub-categories, as we have seen in the natural sciences. It is the films that inspire these concepts, which are still evolving as I continue to study the cinema.

Michel Chion, November 11, 2012

SOURCES

Chion, Michel (1990), *L'Audio-vision: Son et image au cinéma*, Paris: Armand-Colin. English edition, *Audio-vision: Sound on Screen* (1994), translated by Claudia Gorbman, New York: Columbia University Press.

Chion, Michel (1982), *La voix au cinéma*, France: Cahiers du cinéma.

Chion, Michel (2009), *Film, A Sound Art*, translated by Claudia Gorbman, New York: Columbia University Press.

Deleuze, Gilles (1985), *Cinéma 2: L'Image-temps*, Paris: Editions de Minuit.

Fellini, Federico (1976), *Casanova*, film, Italy: Produzioni Europee Associati, et al.

Fellini, Federico (1987), *Intervista*, film, Italy: Aljosha, Cinecittà, Radiotelevisione Italiana, et al.

Gillett, Burt (1931), *The Birthday Party*, film, USA: Walt Disney Productions.

Godard, Jean-Luc (1987), *Soigne ta Droite* (*Keep Your Right Up!*), film, France: Gaumont, JLG Films.

Godard, Jean-Luc (1968), *One Plus One* (*Sympathy for the Devil*), film, UK: Cupid Productions.

Godard, Jean-Luc (1975), *Numéro Deux*, film, France: Anne-age-Bela, Bela Productions, et al.

Godard, Jean-Luc and Anne-Marie Melville (1977–78), *France/tour/détour/deux/enfants*, TV series, France: Institut National de l'Audio-visuel IINA), Sonimage.

Kubrick, Stanley (1968), *2001: A Space Odyssey*, film, USA: MGM.

CONTRIBUTOR'S DETAILS

Michel Chion is a composer of *musique concrète*, a writer, as well as a visiting lecturer at the University of Paris III and various international colleges. Chion has developed the study of sound, particularly the sound of cinema, in several books of which two, *Audio-Vision, Sound on Screen* and *The Voice in Cinema* were published in English by Columbia University Press (translated by Claudia Gorbman). He has also written numerous essays on film directors, and films as well as music. He serves on the editorial board of this journal.

Contact: michel.chion@wanadoo.fr
www.michelchion.com

MICHEL CHION

Le Clap

C'est par un vigoureux coup de "clap" – pour lancer le tournage d'un nouveau film – que se conclut cette célébration du cinéma qu'est l'Intervista de Fellini. Si le clap est une pratique dont peu de gens connaissent la fonction, il n'en est pas moins devenu depuis longtemps un symbole éminent du septième art, où il est l'équivalent de ce que sont pour le théatre les trois coups du "brigadier": la promesse magique du spectacle à venir. Mais avant de devenir un symbole, il servait à quelque chose; le coup sec du "clapman" était destiné notamment à inscrire sur la piste sonore un signal reconnaissable qui, calé au montage avec le moment visible du choc, devait permettre de resynchroniser les supports séparés d'image et de son. Le clap représente donc à lui seul le cinéma sonore, en tant qu'union de ses deux constituants.

La synchronisation audio-visuelle, voilà qui aujourd'hui parait à tout le monde banal, allant misérablement de soi et indigne du moindre intérêt cinématographique, surtout pour ceux qui ont subi le lavage de cerveau de l'idéologie anti-synchroniste des années 70, celle qui honnissait toute correspondance terme à terme du son et de l'image et célébrait, dans les films de Duras, les paroles flottantes et les sons en perdition. Pour elle, le clap était le symbole même de l'esclavage humiliant qui avait jusque là retenu le son dans les chaînes de l'image. Et cependant, le synchronisme n'a pas toujours été perçu de cette façon. Pour les premiers spectateurs du cinéma parlant, dans la fin des années 20, c'était au contraire un phénomène émerveillant. Le synchronisme des films était vanté dans les publicités comme l'est aujourd'hui la fidélité des appareils de reproduction sonore. Il faut donc que les multiples expériences ponctuelles de cinéma parlant présentées antérieurement, par exemple dans les "phonoscènes" de Gaumont montrées à Paris entre 1906 et 1912 environ, aient laissé sous ce rapport à désirer. De surcroît, au moment où se produisit la rupture du parlant, la recherche d'un synchronisme étroit entre le film muet et sa musique d'accompagnement jouée en direct était devenue chez certains une obsédante préoccupation, suscitant l'invention de divers systèmes mécaniques et métronomiques pour mieux diriger la main du chef d'orchestre ou du pianiste, et faire tomber précisément les effets musicaux sur les moments voulus... L'union synchrone du son et de l'image, n'étant pas encore évidente et acquise automatiquement, suscitait, on le voit, un désir aussi passionné que peut le faire, dans une répétition de musique de chambre, la recherche d'un ensemble parfait.

Pourtant, au départ, qu'un son soit entendu synchrone avec l'image d'un événement observable n'est pas un phénomène qui, pour être observé des milliers de fois par jour, aille tellement de soi. Par essence le sonore est devant le visible comme les carabiniers de la chanson: toujours en retard, et cela à double titre. D'une part parce qu'il est souvent un effet, et non une cause, donc venant après, même infinitésimalement. D'autre part à cause de sa lenteur de propagation: 340 mêtres par seconde en milieu aérien, soit presque un million de fois plus lent que la lumière! L'apparent ensemble du son et de l'image ne tient donc qu'à une approximation de notre perception humaine, mais cette approximation est la base même de notre expérience du monde: l'enfant au berceau est d'emblée sensible au synchronisme des phénomènes visibles et audibles; et au cinéma il est avéré, par le principe même du bruitage, qu'un son que l'on fait tomber en même temps qu'un mouvement visuel lui est automatiquement accolé, même si sa matière, sa couleur, son timbre, ne sont que très grossièrement voire pas du tout cohérents avec le phénomène qu'il accompagne. Que ce synchronisme du son et de l'image, dont le clap est le symbole, ait été pour le cinéma sonore tout neuf un phénomène encore magique, rencontre féérique de deux principes pour produire l'étincelle de la vie, cela est avéré par certains des tout premiers dessins animés de Walt Disney, au début des années 30, ces étonnantes Silly Symphonies ("symphonies foldingues") que parfois – trop rarement – diffuse la télévision. Il s'agit en effet de courts-métrages musicaux consacrés à de délicieux concerts miniatures donnés par des souris, canards, et autres animaux, sur des instruments extravagants qui parfois deviennent vivants, s'animent à leur tour, se mettent à jouer d'eux-mêmes une musiquette déchainée dans un synchronisme audio-visuel absolu. Et pour cause, puisque chaque image a été dessinée pour corre-spondre à un son enregistré au préalable, suivant une technique fréquente dans le cinéma d'animation.

Qu'un cinéma sonore tout neuf privilégie le spectacle du concert est logique: rien ne parait plus apte à illustrer la loi du synchronisme que le phénomène instrumental, lequel est par excellence un égrènement de micro-actions, de petits chocs visibles donc chacun est destiné à produire un son distinct. Pas étonnant donc si l'épopée du synchronisme, le cinéma d'animation sonore à son aube n'a pu trouver de meilleur moyen de la chanter qu'en montrant des concerts animés. Voici par exemple *The Birthday Party*, de 1930, évocation farfelue d'une fête d'anniversaire de Mickey Mouse, qui est aussi une fête de la synchronisation. Les amis animaux de Mickey lui ont offert un piano droit, dont aussitôt il joue en virtuose, de ses célèbres mains gantées à quatre doigts. Puis Minnie s'installe à un autre piano, et nos souriceaux de jouer dos à dos un duo frénétique qui ferait paraître poussifs en comparaison ceux des soeurs Labègue. On voit aussi dans ce film, entre autres fantaisies, un métallophone s'animer sur ses quatre pieds comme un cheval, et jouer de lui-même en faisant retomber sur son dos les petits lames inégales, juste comme il faut pour produire, synchronisée note à note, une musiquette endiablée. Le parallélisme du son et de l'image est ici assemblé avec amour comme les rouages d'une montre merveilleuse, dans le ravissement neuf de la synchronisation.

Ce qui est amusant, finalement, c'est que ce symbole dynamique du clap, union énergétique du son et de l'image, ce soit Fellini qui nous le ait offert, lui qui plus que tout autre au cinéma a usé d'un poétique flottement, d'un nébuleux à-peu-près dans l'accrochage des sons aux images. En bon cinéaste italien, il double en effet tous ses films (pas de son direct du tournage) mais au doublage laisse subsister exprès, entre le mouvement des bouches et l'articulation des voix "post-synchronisées", une marge, une indiscipline légère. Rien à voir cependant avec les voix désincarnées, plus off que l'off, des derniers films de Marguerite Duras. Chez Fellini les voix suivent bien les corps, mais avec un léger cahotement. Ce n'est que dans Casanova qu'il s'est permis de pousser le désynchronisme à l'extrême, puisqu'il y arrivait à la voix doublée du séducteur (celle, pour la V.O. italienne, de l'acteur Luigi Proietti) d'être en décalage de plusieurs secondes sur le mouvement visible des lèvres de l'interprète, le Canadien Donald Sutherland. Même là cependant, on restait dans un cinéma du synchronisme; simplement le film paraissait se dérouler dans un milieu différent de celui de l'air, glauque et onirique, où le son anticipait ou retardait très sensiblement sur la vision . . .

Au fond, il y aurait, par rapport à la question du son et de l'image, deux positions, je parle pour ceux qui repensent de fond en comble les moyens de leur art, de Welles à Tarkovski en passant par Bresson ou Tati. Il y a d'un côté les "dualistes" qui pensent le son et l'image comme un couple, comme étant fondamentalement deux, et donc ne conçoivent entre eux que haine ou amour, union ou séparation également absolues: Godard, Duras, et plus dans le passé, Pagnol, Renoir, sont de ceux-là. Et d'autre part, ceux qui les voient simplement comme deux pôles créant par leur rapprochement quelque chose d'autre que du son et de l'image, d'irréductible à l'un ou à l'autre, ou à l'un et l'autre: un effet, une énergie, un courant, une sensation. Fellini est par excellence des ces "triangulaires", que j'oppose aux dualistes. Dans le même temps où sortait le Fellini, un autre hommage au cinéma d'un réalisateur vedette était dans les salles: celui de Godard avec Soigne ta droite. Déjà le titre, déjà l'affiche (deux mains dont l'une serre une autre, laquelle est immobilisée par des menottes) nous disent tout de suite le dualisme de l'auteur. Celui-là même qui a réalisé des films nommés One plus one, Numéro deux, ou encore: France tour détour deux enfants, traite ici encore une fois le son et l'image l'un par rapport à l'autre comme un couple conflictuel. Son et image s'y poursuivent à toute allure, chacun enfermé dans son être. Chaque effet, brillantissime, de montage son-image, de décalage, de retard ou d'anticipation du son par rapport à l'image ou réciproquement, y est comme un coup que l'un donne à l'autre; comme un point gagné ou perdu, toujours provisoirement, dans l'étreinte ou dans le combat. Un "clap" chez Godard – et il y en a eu un certain nombre, car il fut l'un des premiers à représenter le cinéma à l'intérieur même des films – ne peut être qu'un coup de gong pour un nouveau tournoi, et non comme chez Fellini l'étincelle d'où sort la vision d'un monde, dont les constituants de son et d'image sont mêlés, transfigurés.

PHILIPPE CIOMPI

Audio meditation for the New World

ABSTRACT

The Uus Maailm (New World) district of Tallinn in Estonia has for a number of years seen a large number of actions initiated by its residents, aimed at reviving this neglected post-socialist neighbourhood, as well as taking part in the capital's city planning. Community work, a residents' library, street festivals, international networking of hyperlocal communities, and other ongoing activities are taking place.[1] Following a film shoot on the Estonian island of Vormsi (with Ben Rivers & Ben Russel) involving a number of New World residents, this paper was written as an invitation by them to reflect on sound in their city.

KEYWORDS

sound environment
sound in urban context
soundscapes
psychology of sound
Estonia
recorded sound
emotional content of sound

1. http://2010.uusmaailm.ee/ eng/ https:// www.face book.com/uusmaailm? ref=hl

Writing about sound – something that you can't see and which, when reading this paper, you won't be able to hear – are there good words to do that ?

What I really would like to do is to play you a recording of the city of Macau at 6 in the morning in 2004, with the song of caged birds carried by old men, the old ladies chatting happily while doing their Tai Chi exercises and singing Chinese opera songs, with the wind and some sirens, while the Jesuit church bells ring.

The New Soundtrack 3.1 (2013): 27–29
DOI: 10.3366/sound.2013.0033
© Edinburgh University Press
www.euppublishing.com/SOUND

2. http://muniak.com/
   mazen_kerbaj-starry_
   night.mp3

Or the sound of Beirut on that night in 2006 when Mazen Kerbaj, on the roof of his building, improvised on his trumpet while the city was bombed.[2]

Or the rain on the tin roof of the kitchen shed or of the sauna in Vormsi where we and some people from Tallinn's New World district recently spent two beautiful weeks.

When you hear such a recording of a city or a place, you instantly travel in time and space. It happens at once, you don't think about it, but you feel you are in a different place. The sound surrounds you from all sides and brings you to an unknown universe, on some mountain, in a country train station with distant horns and weird birds, near the sea or to New York.

Those sounds also resonate in our memories, measuring themselves against the auditory vocabulary that we have been collecting ever since we're born and throughout our lives. An audio cue brings back a winter street of our hometown with the dogs barking, we remember the voice of our grandmother before she died or a fabulous party when we were drunk.

And the sound environment also touches us physically − our body relaxes, becomes tense, we want to cry, we feel happy, in love or sad.

We live in a constant soundtrack of noises, voices, radios, birds, winds, cars, and there is no way to block sound from reaching us.

Navigating this sea, we learn through our ears. We know if there is a truck coming behind us that we need to beware of, that the children are making a mess in their room, that the neighbour is coming back from work at 2am, slamming the door.

Just close your eyes, when you are in a street, or right now. How many things can you hear at the same time? Someone inside their house? A siren from the city? A clattering seagull on a roof? The old guy cleaning the street with his stupid leaf blower?

Because we are so used to orienting ourselves with our eyes, we assume that our knowledge of the world around us happens fundamentally through vision. But interestingly, the practice of a film sound designer, and to a point that of a sound artist, is rooted in this parallel understanding that a large amount of information about our environment comes to us through our ears.

Only sound works at this intriguing level, not really conscious, and not totally unconscious either. We process the information it conveys, we get the cues coming from the soundscape, but we don't realise that we are doing it. Without further questionning, we naturally integrate them into our behavior. Often, even, we are convinced that we have seen something, when we have only heard it.

I like to explore all those layers, near or distant, loud or quiet, that come to us through our ears, and discretely inform us about the city, the time, the pace of events.

When I replace the natural soundscape around us with another, recorded one, the same thing happens: we recognize and reconstruct a new, complex and complete environment, and we get this strong sense of being in that place at that time. Much more so than if we look at an image of that place. With vision, we recognise the place, with sound we enter it.

Sound takes us in its arms, its whirlpool, its cacophony, its logorrhea, and soothes us like babies or tortures us like Guantanamo inmates. . .

When there is more silence we can hear further into the world. At night you can hear people walking in the next street, a late tram, an ambulance. That slowed down, linear timing of sounds changes your mood and your perception. Do you feel the nostalgia of the night? Or maybe you feel tired and want to go to sleep then?

I won't. I will go to my sound library and create the rich silent night of my desire – a very quiet hum of the city, with some distant cars and lost motorbikes, two or three different winds and airs, a solitary night cricket, a distant dog bark, a wind swell, some rattling metal moving in the breeze. Maybe we can hear a few swells and clanks of the harbour in the distance, or a Chinese train whistle reaching us through the plains and echos? I will create a suspended time in an imaginary space and I will invite you to enter it with your ear. Will you come?

CONTRIBUTOR'S DETAILS

Philippe Ciompi is a film sound designer with a long lasting interest in sound art, free improvised music, ethnomusicology, electronica and more. His collaborations with directors such as Andrew Kötting, Stephen Dwoskin, Xiaolu Guo, Ben Rivers, the Quay brothers, as well as his sound art work and film scores, have led to a specific and particular body of work often concerned with sound textures, sound space and time articulation. He also teaches sound design at the Royal College of Art - London and at HEAD – Geneva.

Contact: perspectivefilms@gmail.com

JOHAN-MAGNUS ELVEMO

Spatial perception and diegesis in multi-channel surround cinema

ABSTRACT

This article addresses the challenges that are connected to the current use of the concepts 'diegetic' and 'non-diegetic', when sound cinema may reach an audience from many channels and directions. It poses the question: 'How can we relate to the division between diegetic and non-diegetic sound within a challenging multi-channel digital surround world, where cinema sound totally envelops the audience?' The article argues that the transition area between our perception of the space in front of us and the space behind us is seamless – in a perceptual-cognitive sense. Inside the transitional area between these two fields, there exists a hybrid area, wherein audiences partly support their experience with memory and partly with visual perception. This hybrid area links together our experience of the space in front of us, and our experience of the space behind us. In the movie theatre this condition is most likely to affect the ability of the audience to integrate sound from the surround speakers into the diegetic space of films. The 'exit door' effect, and other often dysfunctional surround effects, are most likely products of situations where the audience is exposed to spatial conflicts due to difficulties with the process of linking together these two fields.

KEYWORDS

multi-channel
surround sound
enveloping cinematic
space
egocentric space
representation
allocentric space
representation

The New Soundtrack 3.1 (2013): 31–44
DOI: 10.3366/sound.2013.0034
© Edinburgh University Press
www.euppublishing.com/SOUND

Since the 1950s, the terms diegetic and non-diegetic have been much used tools in the discussion of how sound and music are linked to cinematic space. Diegetic sound has been defined as all sounds that are generated by actions and events within the cinematic universe: in other words, sounds that the characters in a film should be able to hear. Sounds that are generated outside of this space, as, for instance, film music and voice over, that the characters in the cinematic universe should be unable to hear, are defined as non-diegetic.

The distinction between the two categories has largely been regarded as contained within the cinematic object itself, the film, and explained as a classification process, integral to the filmic experience.

The terms diegetic/non-diegetic have been proven to be useful tools in the analysis of the relationship between sound objects and cinematic space, when dealing with films that have standard mono or stereo sound tracks. However, they are not equally suited to describe the relationship between different kinds of surround sound objects and cinematic space.

The point of departure for this article is based on the following question: How can we relate to the division between diegetic and non-diegetic sound in a challenging multi-channel digital surround world, where sound and music totally envelop the audience in the movie theatre?

The main aim of this article is to show that the perceptual experiences in surround sound movie theatres (as well as in private home theatres that are equipped with surround speakers) together with the different technical screening conditions that we may experience in these situations, easily affect the process whereby audiences classify different surround sound objects as integral diegetic parts of the cinematic spatial universe.

ENVELOPING CINEMATIC SPACE

Many writers describe with great enthusiasm how digital surround systems today offer filmmakers the possibility to envelop the audience with sound. It is often claimed that the diegetic film space has expanded beyond old borders. Michel Chion, for instance, in his book *Audio-Vision* (1994: 84), uses the concept 'absolute off-screen space', to describe the effect of surround sound in cinema.

In 'Enlarging the Diegetic Space: Uses of the Multi-channel Sound-track in Cinematic Narrative', Manolas and Pauletto claim: 'Today's multi-channel soundtrack ... is able to enlarge the diegetic space and immerse the audience in it'. (Manolas and Pauletto 2009:39).

In *Beyond Dolby (Stereo), Cinema in the Digital Sound Age,* Mark Kerins claims, in a similar way, that the use of surround sound fundamentally changes the film experience:

[T]he idea is that the audience is literally placed in the dramatic space of the movie, shifting the conception of cinema from something 'to be watched from outside' – with the audience members taking in a scene in front of them – to something 'to be personally experienced' – with audience members literally placed in the middle of the diegetic environment'. (Kerins 2011: 130)

Such a personal experience of being immersed into the middle of a cinematic diegetic environment may conflict with the fact that we, as audiences, are also aware that we sit in a real movie theatre. In his book *Sound Designs & Science Fiction*, William Whittington explains that 'the use of surround offers a total sonic environment, which masks the real environment of the theatre space to create a sonic space with no entry and no exit.' (Whittington 2007: 122).

In other words, surround sound should contribute to the audience's experience of diegetic space in cinema, in a manner that allows them to experience this space in much the same way as the characters in the film do.

Many writers (and viewers) believe that surround sound has the ability to change completely the film experience, as it is normally defined, by amplifying its power and expressiveness. However it is also a true that surround sound may not work in this idealised way. Very often sound objects emitting from the back and from the sides of the movie theatre are experienced both as distracting and disturbing.

One example of this kind of dysfunctional experience that surround sound can produce is called the 'exit door' effect. When this occurs, we experience that a sound emanating from behind us is not rooted in the diegetic universe of the film. It is generated in the space of the cinema theatre itself, by a specific loudspeaker, which we are being made conscious of by the effect. In a case like this, the sound from behind will be experienced as neither diegetic nor non-diegetic – it would simply be regarded as foreign to the cinematic expression entirely.

Another, often dysfunctional use of surround sound is the 'in the wings' effect, where the sound of a character or an object visually entering or exiting the screen, sonically approaches the front or retreats to the rear of the theatre. Very often, this kind of sound is not properly integrated into the diegetic universe. Both these dysfunctional effects are most likely products of spatial perceptual conflicts.

VISUAL PERCEPTION OF SPACE IN THE MOVIE THEATRE – THE TWO SPACES

In the cinema, as in real life, the audience has a field of view that is approximately 180 degrees horizontally, and 100 degrees vertically. The frame of the screen defines the limits of the visual cinematic expression. In most cinemas the frame does not expand beyond the limits of our own visual field. What is outside the frame, but still part of our field of view, belongs to the space of the movie theatre itself.

Thus, when seated in the movie theatre, we are confronted by, or placed in, two different types of visual space or representations of such space. One is the space of the movie that we are watching, which can be divided into two further kinds of spaces - onscreen and offscreen - as Nöel Burch described in his book *Theory of Film Practice* (Burch 1973). The other is the space of the movie theatre itself. The former lacks any visual clues about what might be behind the cameras point of view, since one spatial 'wall' is missing at all times. It is only with the help of memory and other factors such as sound, in combination with the editing of the different shots and point of views building the scene, that the viewer is able to construct some kind of cognitive representation of this part of the diegetic space, as well.

The latter – that is, the space of movie theatre itself, is also divided in two parts. One is *visual* and in front of us, the other *out of vision* and behind us. Although the occluded parts of the space of the cinema theatre are non-visible, they are conceptually present for us due to our memory and a variety of perceptual cues.

AUDITORY PERCEPTION IN THE MOVIE THEATRE – THE TWO SPACES

In the discussion of auditory fields in the book *Listening and Voice: Phenomenologies of Sound*, Don Ihde describes the term 'field' as a 'limited and bounded context.' With this he means, '... that all things or occurrences are presented in a situated context, "surrounded" by other things and an expanse of phenomena within which the focused-on things or occurrences are noted.' (Ihde 2007: 73).

In other words, every sound that we attend to is placed – by us as listeners - into some kind of spatial relationship with other sounds, objects and happenings in the environment.

The auditory world of the movie theatre can also be characterized by the two spaces mentioned above. That is: the cinematic space that any film represents, and the space of the movie theatre itself.

In surround cinema, these two different spaces are represented both in front of as well as behind us. Sorting out the 'situated context' of sounds in front of us is normally not so complicated, since we can support our auditory perception with some kind of visual cues. Sound objects from behind, deprived of any visual support, need to be processed solely on the basis of their auditory qualities. Such processing is usually more demanding for the audience, as sounds easily may be placed in the wrong 'bounded contexts', as is the case when the above described exit door effect occurs.

THE TWO COMPETING SPACES BEHIND US

It is a fact that we always have a sense of a real space behind us - including those moments when we are seated in the movie theatre. Although it is also obvious that the main goal of surround sound cinema is to create an experience of a diegetic cinematic space, which also is behind us. How we as audiences react, behave or adjust - perceptually and cognitively - to these two competing 'back' spaces is a challenging question, and one that needs to be addressed within the discipline of film studies.

If audiences are able to experience an enveloping diegetic cinematic space, this requires the creation and perception of a diegetic cinematic space behind them, in a similar way that we all experience the out-of-vision parts of our real life spaces. The question is: Could such an illusion of a fictional space behind the audience co-exist with a sustained representation of the non-visual part of the movie theatre, which is real, and behind them at all times? Or, as Whittington (2007) suggests, do we need to mask the real environment of the movie theatre behind us, in order to experience an enveloping diegetic cinema space?

The experience of real visual space in front of us seems to disallow a simultaneous on-going processing of any imaginative visual space, since the process of active mental construction of imaginative visual space will use resources that normally are involved in the processing of information within the visual field. This is why we often need to close our eyes, or look away, when we perform detailed mental constructions of imaginative visual spaces. In the same way, it is likely that it would be difficult, if not impossible, to simultaneously have the two types of space processing connected with the space behind us, in the theatre. One 'limited and bounded context', to use Don Ihde's terminology, will most likely mask the other, or, at least, place them in some kind of competitive relationship where the audience will have to perform a kind of switching between the two types of spaces.

In order to integrate the viewer in to an enveloping diegetic cinematic space, we will have to create a condition in the movie theatre that allows the viewer to temporary shut down most of the processing connected to the real space behind.

The question becomes: Under what conditions might we allow a shut down of the processing of the real space behind us? And under what conditions might the viewer oppose such a shut down?

Although an audience may feel secure in the movie theatre, any shut down of vital parts of our sensual/perceptual apparatus that are directed backwards will under most conditions be met with resistance. Such unwillingness might be seen as protective, in the sense that it aims to maintain the audience's position as anchored in the movie theatre, and as such, separated and detached from the diegetic universe of the film.

It is possible that the emotional intensity of any given scene might affect the degree of resistance that the audience generates towards shutting down the processing of the real theatre space behind it. It is well known that the emotional content of sensory stimuli may affect perception in several ways. In this case, the emotional content of a scene might be one of the factors that have an effect on whether sounds are perceived as diegetic or non-diegetic.

Audiences may also become so accustomed to surround sound cinema conventions and experiences that they learn how to cope with them, and thus more easily accept a temporary shut down of the processing of the real space behind. Mark Kerins seems to be of this opinion, when he complains, 'With digital surround sound now nearly two decades old, audiences should have grown accustomed to movie soundtracks with spot effects in the surrounds, and the exit door effect should no longer be a concern.' (Kerins 2011:159).

The fact that the dysfunctional effects mentioned above ('exit-door', 'in the wings'), still seem to create problems, even two decades after the introduction of surround sound, indicates that it is not necessarily true that we can be trained to integrate all kinds of surround sounds emitting from behind.

THE EXPERIENCE OF REAL SPACE BEHIND US

In 'Phenomenal Awareness of the Surrounding Space: An Ecological Perspective', the authors state:

> Traditional psychology has dealt with only a fraction of spatial perception, that is, the acquisition of definite information from within the visual field. However, such perception, especially under natural conditions, is always embedded in the indefinite awareness of the entire surrounding space. It is necessary to recognize the importance of this awareness and reconsider the traditional concept of visual perception in order to elucidate the nature of spatial perception. (Inagami, Ohno and Tsujiuchi 2008: 141)

A statement such as the one above, and the use of the term 'embedded' in this context, indicates that it is not necessarily the perception of what might be in our visual field at any given moment that constitutes the most important part of our experience of space. On the contrary, it might be our perception of what is outside of our visual field that contributes most to our experience of space.

In his book *Phenomenology of Perception*, Merleau-Ponty describes the experience of the space behind:

> The region surrounding the visual field is not easy to describe, but what is certain is that it is neither black nor grey. There occurs here an indeterminate vision, a vision or other, and, to take the extreme case, what is behind my back is not without some element of visual presence. (Merleau-Ponty 2004: 6)

The term 'indeterminate vision' focuses primarily on a perceptual shortage connected with the sensing of the space behind. It is obvious that we are not able to see what is behind us, but our sense of the space behind us is undoubtedly both strong and present at all times and, not least, our sense of the space behind us does not interfere with our constant visual processing of the space in front of us. The quasi-visual presence of the space behind us, which Merleau-Ponty describes, is in this respect undoubtedly very functional.

Other scientists and writers have also tried to explain and establish models for the way we experience the space behind us. In his article *Gestalt Isomorphism and the Quantification of Spatial Perception*, Steven Lehar describes our perception of the space behind us in this way:

> Parts of the visual world that are currently outside of the visual field are experienced amodally, i.e., in the absence of a vivid impression of colour and visual detail. However the world behind the head is experienced as a spatial structure ... (Lehar 1999, Para 28)

In 'The Primacy of the body, not the primacy of perception', E.T. Gendlin questions whether our experience of the space behind our backs can be defined as perception at all, and reflects that 'it is not vision, hearing, or touch, nor is it just the togetherness of the five senses. It is rather a direct bodily sense that you have and use all the time.' (Gendlin 1992: 346).

In light of these and other works, we might summarise that the space behind us, as we experience it in the movie theatre and elsewhere, is experienced as amodal spatial structural sensing. Such can be described as a sensation of a space containing relationships and ideas; for example, larger or smaller areas, openness, closed-ness, relative distances, meeting places, such as corners, lines, and so on. The objects in this amodal structural space are more conceptual than they are concrete. In other words, we relate to the idea of a chair, the idea of a table, and so on, not to a specific chair or a specific table.

It is this illusion of the structural and conceptual sensing of the space behind us that the surround sound should try to establish.

ALLOCENTRIC SCENE BASED AND EGOCENTRIC SPACE REPRESENTATION

Recent research indicates that humans encode visual information about the world in two distinct ways. In *Sight Unseen,* Goodale and Milner describe these two different visual systems, and the way the brain divides visual processing according to what 'job' needs to be done at a given moment. The two visual systems (or streams, in the ventral and dorsal cortex) have different responsibilities. The ventral stream utilises scene-based frames of reference. That is, visual information in this stream 'puts objects in their context. We perceive the size, location, and motion of an object almost entirely in relation to other objects and surfaces in the scene'. (Goodale and Milner 2004: 74).

Tatler and Land, in the article 'Vision and the representation of the surroundings in spatial memory', use the term 'allocentric space representation', when they refer to what Goodale and Milner call scene-based frame of reference, and they define this way of representing space as 'map-like ... independent of our current location and heading and survives over extended periods of time. This representation must ... be built up from vision over time, but does not rely on immediate visual input'. (Tatler and Land 2011: 605).

This type of spatial representation is also active when we remember the spatial layout of a town, or the layout of an apartment in a television series that we have followed for some time. When the doorbell rings off screen in a known cinematic location, we know where the entrance door is within that cinematic space. A scene-based allocentric frame of reference makes it possible for us to build a representation of a scene without going through an absolute, precise and costly processing of all the exact distances between and sizes of the different objects.

The dorsal visual stream, on the other hand, is in charge of the part of visual perception that has as its main responsibility to prepare the body for action. This system cannot rely on relative distances and sizes but needs absolute data about the surrounding world. This data requires, according to Goodale and Milner (2004), an egocentric frame of reference in order to work properly.

The egocentric representation of space is a viewpoint-dependent temporary representation based on the placement of objects and happenings relative to our own body position and our heading in the environment that surrounds us. When we speak about left and right in connection to the space in which we are placed, this usually is a product of our egocentric representation of the space. In the cinema we often have other moviegoers on each side of us, and, now and then, a tall person in front of us, which makes it difficult for us to see the entire screen; we might also feel when the feet of the moviegoer behind us touches our seat, and so on. Although sitting in the darkness, at the same time we remember how the movie theatre itself is constructed; recalling the location of fire exits, toilets and other landmarks. Such factors define our sense of egocentric placement in the space of the movie theatre. To sum up, we might say that we process the part of space that is 'out of reach' allocentric, and the space that is 'within our reach' egocentric.

Whether surround sound objects are experienced as diegetic or non-diegetic within the spatial experiences of the movie theatre can be discussed in light of these two terms, and may thus prove to be useful tools in the further exploration of the challenges connected to both surround sound production and perception.

ALLOCENTRIC AND ILLUSIVE EGOCENTRIC SPACE REPRESENTATION OF DIEGETIC CINEMATIC SPACE

The ways in which different cinematic spaces are represented are different from the ways that real life spaces are represented. When watching movies, audiences usually create a form of spatial relationship between the geographical landmarks in the diegetic universe. If the filmmaker follows the rules of cinematic continuity, audiences are pretty good at establishing map-like representations of the different environments within a film. This kind of representation of space is firstly allocentric and scene-based. That is, the viewer remembers the relationship between different objects and landmarks in the diegetic space, and is able to orient themselves in the cinematic environment by the use of this knowledge, but they have no experience of any of the absolute aspects of the different spatial parameters that are represented in the scene. A change of lenses, for instance, may alter the relative perceived distance between objects in the scene, but without this disturbing the audience.

Although audiences may have no strong sense of being directly and egocentrically placed in a cinematic space, it might be argued that their perception of this space at least resembles some of the qualities associated with egocentric space perception. One of the keys to this argument lies in the fact that cinematic objects are occluded in the same way that real life objects are occluded. That is, the placement of a camera cannot avoid representing several point-of-view-dependent occlusions of a scene.

A camera movement, for instance, can be seen as a dynamically occlusive change, that adds depth sensations to a scene.

Any cinematic occlusions might be seen as an embodiment of an imaginative viewer. A cinematic occlusion always represents a position of a viewer in space, in the same way that point of view dependent occlusions do in real life. It is therefore possible to argue that the occluded parts of any given cinematic image may contribute to the creation of an illusive egocentric encoding of the cinematic space itself. An illusive egocentric encoding of the cinematic space could then, in combination with the emotional content and experience level of the audience, be one of the factors that contribute to a temporary shut down of the processing of the real space behind us.

This possible illusive egocentric encoding of a visual cinematic space will, however, necessarily be restricted to the screen area and the spatial area close to the borders of the screen. Although we usually are able to construct some amodal ideas about the offscreen space in relation to the scenes that we are watching, in most cases we are not able to go far beyond the border of the screen in this process of space construction. Because of this restriction, it is most likely that there always will be some kind of spatial gap between the borders of the illusive cognitively constructed diegetic offscreen space in front of us, and our possible experience of a diegetic space behind us, which mainly must be a product of auditory spatial cues from behind, along with some overall cognitive narrative understanding of the spatial qualities of the fictional universe. The auditory spatial cues from behind - or from the sides of the theatre - and the knowledge and the understanding of the diegetic universe acquired through the watching of the film, are probably not enough to bridge the gap between our experience of a diegetic egocentric space behind us, and the restricted diegetic on- and offscreen space in front of us. Thus it might be argued, that it is not possible to establish a complete diegetic enveloping space. The diegetic enveloping space will, at best, always be split in some way or another.

'IN VISION – OUT OF VISION' – THE PERIPHERAL VISUAL FIELD

The spatial experience when seated in the movie theatre is first and foremost egocentric. That is, we are perceptually placed, more or less enveloped in darkness, in the middle of our own sensory world, where the screen is to our front. Since we normally are seated and deprived of the possibility to move around within the space, the spatial situation remains relatively static, but it is divided between what is inside or outside the visual.

The visual field is not homogeneous. Our vision of depth, for instance, covers approximately only 120 degrees of the 180 degrees that define our total visual field. On each border of the visual field, an approximate 30 degrees periphery, that only receives information from each eye separately, has an effect on the way depth is perceived. The perception of colour and close detail decreases as we move towards the periphery of the visual field.

In other words, as an object occurs more in the peripheral borders of our visual field, our visual perception of it is gradually decreased. Importantly, this area of decreased vision defines the borders to our experience of the out-of-vision space behind.

This transitional area between our perception of the space in front of us and that of the space behind us is seamless, in a perceptual-cognitive sense, regardless of the fact that we experience each space in different ways. The peripheral part of our field of view is characterized as having less information regarding depth, colour and detail, than is our central vision. This peripheral view creates a natural transition to the space behind us, that obviously lacks these qualities altogether.

Within the transitional zone between these two fields, we find a peculiar hybrid area, wherein our experience is partly supported by memory and partly by our visual perception.

> It is usually assumed that we use vision to locate objects that are within our peripheral field of view, but there is evidence that this is not always the case, and that memory may be equally important even for objects that are plainly visible. (Tatler and Land 2011: 603)

That is, we can relate to objects that are entirely visible within the periphery of the visual field as much by using our memory as by our sense of sight. This means that we do not necessarily 'see' objects that are actually visible in the periphery, but utilise memory to define the positioning of such objects in space. This relationship between our experience of the space behind us and the space in front of us is updated dynamically, as we move our bodies and head in space. The result is that we generally have good control, not only of what we see, but also of what lies behind us. Our experience of the unseen space behind us plays an important role in the egocentric experience of that space.

THE AUDITORY FIELD AND THE SPACE BEHIND US

There is a tendency to regard the auditory field as more perceptually homogenous than it actually is. In 'When Room Size Matters: Acoustic Influences on Emotional Responses to Sounds', Ana Tajadura-Jiménez and her co-authors describe how natural sound objects produced in the space behind our visual field have a tendency to affect us emotionally more powerfully than sound objects that are generated within the borders of the visual field:

> Results confirmed that sounds from 'back' locations, especially natural ones, led to an increase in subjective and physiological arousal. These results might suggest the existence of an auditory attention and emotion bias towards the space outside ones visual field. (Tajadura-Jimeénez et al., 2010: 14)

To a certain extent sound from behind the visual field is perceived differently from sound originating within our visual field. In some instances sound from the space behind can be given prominence in the overall perception of an audio expression.

In 'From surround to true 3D', Lennox, Myatt and Vaughan describe some of the criteria that we use when we pick out sound objects for special processing, that is, we give them a 'where' and a 'what' processing. They use the term 'perceptual significance' to describe how we select for attention those properties that have the potential to supply us with information that we might need (Lennox/Myatt/Vaughan 1999). For instance, new sounds will receive more attention than familiar sounds; strong sounds that stand out will receive more attention than weaker sounds that blend in. Sounds which indicate that they are approaching will receive more attention than sound which are receding. Sounds that indicate the source of the sound is approaching us closely receive more attention than sounds that are indicating that the source of the sound is approaching us from afar.

In 'An Adaptive Bias in the Perception of Looming Auditory motion', John G. Neuhoff describes experiments that confirm that sounds that approach listeners are perceived as stopping closer to the listeners than sounds that recede, despite the fact that both types of sounds have identical stopping points.

> A bias for looming auditory motion may provide a selective advantage in preparing for contact with an approaching source, or an increased margin of safety on approach. If the source is perceived as closer than it actually is, then the listener will have longer than expected to prepare for the source's arrival. (Neuhoff 2001: 100)

These findings indicate that some types of sounds are integrated more easily in an egocentric space perception than are others, since they seem to trigger a 'preparation for action' directed at the real space that envelopes us.

If at the same time our representation of the diegetic cinematic space in front of us is mainly allocentric, it is likely that the resultant effect is that of a divided diegetic space. That is, the space in front of us is represented mainly as allocentric and that behind us is represented as mainly ego-centric. It is possible that it is the conflict between these two ways of representing the diegetic space that results in a mistaken anchoring of the intended diegetic surround sound from behind.

WHAT IS THE SOLUTION?

The conflicts mentioned above are common in cinema. The main challenge for a sound designer who wants to integrate surround sounds into the overall auditory expression without breaking the diegesis, is to create a sound design expression which can emanate from behind and from the sides of the auditorium, a design that does not require the audience to create an active concretisation or imagination of the fictive space behind, but which nevertheless represents that space.

In itself, this is a more difficult task than one might believe. Our auditory perceptual apparatus has as one of its main functions to identify the sources of the sounds we hear. Determining the placement of sound sources in space is therefore very natural for us. This act of interpretation and mental placement of sound objects in space demands both a 'where' and a 'what' processing of the auditory expression. Both types of processing present a risk of the breakdown of the relation between these sounds and the diegetic space.

In his description of what he calls 'ambient sounds', Michel Chion points to the case that ambient sounds are sounds that are not normally processed completely. 'Let us call ambient sound sound that envelops a scene and inhabits its space, without raising the question of the identification or visual embodiment of its source: birds singing, church bells ringing.' (Chion 1994: 75).

These types of sounds are relatively easily integrated into the diegetic film space since they don't form perceptual foregrounds in the overall cinematic expression. It is unlikely that we would submit the sound of rain to a 'where' processing, similarly the sound of wind in the trees, or the sounds of ricochets in a battle scene. Sounds which come from the back or sides of a movie theatre and are meant to represent the diegetic space should therefore be of such a character that the audience does not require specific spatial processing. It is a paradox that it is the sound from behind that is not spatially processed by the audience, which creates the illusion of an enveloping diegetic film space. Specific sounds from behind that demand spatial processing destroy the illusion and bring the audience out of the diegetic space and back into the movie theatre.

If filmmakers and sound designers take into account these and other factors that might affect how we direct our attention towards auditory objects, many of the dysfunctional perceptual/cognitive conflicts that audiences experience in the movie theatre could be avoided.

Some factors, though, may be out of reach even for filmmakers. The placement of people in the audience in the movie theatre, for instance, may in itself affect whether sounds are integrated or not in the diegetic space. Dolby uses the terms 'sweet spots' and 'surround zones' to describe the parts of the movie theatre where the surround sound is best experienced. In some cases, when sitting towards the sides of the movie theatre, the effect of distinct sounds from the side speakers may vary a great deal compared to the situation that the audience in the middle of the movie theatre experiences – for instance, how close or how far the sounds are perceived to be, or how strong or weak they might sound. In a similar way, being seated in the front or back of the cinema may affect perception. Equally, the quality of the technical sound systems in the movie theatres, and the way these systems are adjusted, can affect the integration or non-integration of sounds in the diegetic space of the films.

CONCLUSION

The technological conditions for media today are constantly changing. This is also the case for the sound systems used in cinemas. At the moment it is common to use digital surround systems that allow 8 separate channels (Dolby 7.1), but screening systems are already being developed that offer

22 separate sound channels (Sony 'Super Hi-Vision', 7680×4320, Audio 22.2). These days, sound does not necessarily only reach us from the front, the sides or the back of the movie theatre, but may also come from above – as well as from below. The slogan of the day seems to be: 'Any Sound, Anywhere'. (Kerins 2011: 69).

When the basic technological conditions in a medium are changed, as for instance when Digital Dolby Surround was introduced, theoretical discussions reflect this change. This means that we may no longer regard the question of whether sounds are diegetic or non-diegetic not only as a product of qualities within the film itself, but also as a result of other factors related to the screening situation and our perceptual experiences of space in the surround movie theatre. To design and create a functional diegetic offscreen space using surround sound is a great challenge for any filmmaker. For us as audience, to be confronted with this type of cinematic expression is also a challenging experience, both perceptually and cognitively.

SOURCES

Burch, Nöel (1973), *Theory of Film Practice*, USA: Preager Publishers.

Burgess, Neil (2006), 'Spatial memory: how egocentric and allocentric combine', *Trends in Cognitive Sciences*, 10(12), pp. 551–557, http://www. sciencedirect.com/science/article/pii/S1364661306002713. Accessed 26 September 2012.

Chion, Michel (1994), *Audio-Vision*, USA: Colombia University Press.

Gendlin, E.T. (1992), 'The Primacy of the body, not the primacy of perception', *Man and World*, 25 (3–4), pp.341–353, http:// www.focusing.org/primacy.html. Accessed 26 September 2012.

Goodale, Melvyn and David Milner (2005), *Sight unseen*, UK: Oxford University Press.

Ihde, Don (2007), *Listening and Voice: Phenomenologies of Sound*, USA: State University of New York Press.

Inagami, Makoto, Ryuzo Ohno and Rieko Tsujiuchi (2008), 'Phenomenal Awareness of the Surrounding Space: An Ecological Perspective' in *Cognitive Studies*, Volume 15(1), pp. 134–143, http://www.jstage.jst. go.jp/article/jcss/15/1/15_1_134/_pdf. Accessed 26 September 2012.

Kerins, Mark (2011), *Beyond Dolby (Stereo) Cinema in the Digital Sound Age*, USA: Indiana University Press.

Lehar, Steven (1999), 'Gestalt Isomorphism and the Quantification of Spatial Perception' in *Gestalt Theory* 21 (2), pp. 122–139, http://cns-alumni.bu.edu/~slehar/webstuff/isomorph/isomorph.html. Accessed 26 September 2012.

Lennox, Peter P., T. Myatt, J. Vaughan (1999), 'From Surround to True 3-D' in the *Proceedings of Audio Engineering Society 16th International Conference, AES, Rovanearme, Finland*, http://decoy.iki.fi/dsound/ ambisonic/motherlode/source/LennoxAes16.pdf. Accessed 26 September 2012.

Manolas, Christos and Sandra Pauletto (2009), 'Enlarging the Diegetic Space: Uses of the Multi-channel Soundtrack in Cinematic

Narrative', in *The Soundtrack*, Volume 2 Number 1, Intellect Ltd, pp. 39–55.

Merleau-Ponty, Maurice (1945/2004), *Phenomenology of Perception*, UK: Routledge.

Neuhoff, Johan G. (2001), 'An Adaptive Bias in the Perception of Looming Auditory Motion' in *Ecological Psychology*, 13(2), pp. 87–110.

Tajadura-Jiménez A, P. Larsson, A. Väljamäe, D. Västfjäll, M. Kleiner (2010), 'When Room Size Matters: Acoustic Influences on Emotional Responses to Sounds', in *Emotion*, 2010 June, 10(3), pp. 416–22, http://psycnet.apa.org/journals/emo/10/3/416.pdf. Accessed 26 September 2012.

Tatler, Benajmin W. & Michael F. Land (2011), 'Vision and the representation of the surroundings in spatial memory', in *Philosophical Transactions, Royal Society, Biological Sciences*, 27, February 2011 vol. 366 no. 1564, pp. 596–610, http://rstb.royalsocietypublishing.org/content/366/1564/596.full. Accessed 26 September 2012.

Whittington, William (2007), *Sound Designs and Science Fiction*, USA: University of Texas Press.

CONTRIBUTOR'S DETAILS

Johan-Magnus Elvemo is Assistant Professor in the Department of Art and Media Studies at the Norwegian University of Science and Technology, Trondheim.

Contact: johan.magnus.elvemo@ntnu.no

ROBERT WALKER

'Don't pump up the emotion': The creation and authorship of a sound world in *The Wire*

ABSTRACT

The use of sound in the HBO TV series, The Wire, *has a variety of distinctive features, which set it apart from other television shows. This article analyses the technological, authorial, logistical and creative factors that have shaped its sound aesthetic. Based upon interviews with the key personnel involved in the show, the show's relationship to genre and cinema is considered, given its use of formal approaches and production methods closer to the cinematic. The role of music and sound effects is considered central to the notions of realism and verisimilitude in the show. The impact of audio delivery formats on how television is consumed has materially affected its content. The influence of 'insiders' and other colla- borators is traced through* The Wire's *soundtrack.*

KEYWORDS

The Wire
verisimilitude
realism
sound effects
sound design
recording
editing
mixing

The extent to which the HBO TV series, *The Wire* (2002–2008), attempts to capture a 'realistic' slice of Baltimore life in and around the city's drug trade is well known. *The Wire* is widely considered to be more in touch

The New Soundtrack 3.1 (2013): 45–59
DOI: 10.3366/sound.2013.0035
© Edinburgh University Press
www.euppublishing.com/SOUND

with the world it portrays than previous television crime drama. The show's approach to sound is an important part of this.

> Verisimilitude was the watchword always and in all things 'Wire'. Many of the events and characters were based on real experiences in David and Ed's [*the show's creators*] combined arsenals. ... It was important to them that the places and situations ring true to those watching who were in the know.
>
> (Ralston 2009)

The consistent sound aesthetic of *The Wire* includes elements of classical Hollywood, cinema verité and documentary technique, and has shaped elements of these traditions in service of verisimilitude. This could be the booming bass of a drug dealer's in-car sound system echoing between the tenements or the chaotic chattering backgrounds of the show's fictional *Hamsterdam* neighbourhood.

The Wire is an important case study in how 21st century television has the potential to find new forms of sonic truth to support its stated authorial goals. Furthermore, the sound production of *The Wire* raises important questions of authorship and authenticity that warrant further examination. This article examines the defining aspects of the show's approach to sound design across 60 episodes. The factors which exert an influence on the sound world of *The Wire* could be classified as aesthetic, technological and authorial. This article draws on comments from the show's creator, David Simon, as well as from interviews I conducted with two key members of the sound team, Jen Ralston (Supervising Sound Editor) and Andy Kris (Re-Recording Mixer). *The Wire* has the potential to find 'the nuance of daily life' (Simon 2004) in its sound world, but as will be demonstrated it takes a lot of effort, technique and artifice to achieve this goal.

THE SPACE THE MUSIC VACATED

In the case of the show's use of music, there are some clearly stated authorial rules set out from episode one by its creator, David Simon:

> ... we don't cue music when there's something exciting happening ... we do not use musical cues ... we don't pump up the emotion by using koto drums or synthesizer, that suggestive stuff that tells you what you're supposed to think. In fact we spent a lot of time on the sound package but what we tend to do is use background noise. You hear police radios in this scene; you'll hear shouts in background ... we fill up the space by trying to create a real ambient sense of the moment ... music has to be source, it has to come from a car radio or a tenement window. We're not interested in montage except at the very end of the piece ... we would break that wall and allow a moment of montage. Other than that we avoid it studiously.
>
> (Simon 2004)

David Simon's description of the approach to music highlights a number of key issues. *The Wire* almost exclusively restricts the soundtrack to representing events as experienced by the characters, not directly addressing the audience through non-diegetic music. This approach to music forms part of establishing an overall aesthetic which does not readily break out of the world those characters inhabit. The soundtrack is largely defined by the characters' hearing abilities and perceptual focus, which determine its content, volume, direction and acoustic perspective. The show only lets you hear what the characters would hear. Without non-diegetic musical score, the show does not necessarily bring with it clearly labeled emotional direction. As Andy Kris comments:

> I would kill to do a show again that didn't use a score to tell the viewer how to feel. I think when people have an emotional response to the show, it's because they formulated their own feelings, without even knowing it. A score can be a very powerful and manipulative tool.
>
> (Kris 2009)

It is clear that directly communicating emotion is more straightforward with music than sound effects. However, there is an important distinction when sound effects occupy the space left by music. These sounds have multiple independent causal relationships and sources (the drivers in the cars of distant traffic are not aware of the police radio which has no bearing on the barking dog), which makes them seem as though they are unstructured. Although these sounds may have been purposefully designed to create a sense of the moment, a mood or an emotion, their authorial intent is disguised because they are plausibly sited in the onscreen world. By contrast, the musical montage sequences, at the end of each season, stand out in sharp relief, removed as they are from the diegesis. This serves to remind the audience that this is a fiction, controlled from outside of the story world by its authors. It could also be argued that the use of montage is very economical in terms of screen time in tying together multiple disparate narrative threads.

The other point Simon raises is that there is a 'space' vacated by the non-existent score that needs to be filled. *The Wire* is by no means unique in eschewing musical score: Hobson (2003) discussing the lack of music in UK soap opera, sees the decision not to include musical score as a 'crucial element both in the way that audiences perceive programmes and in any debates about realism' (Hobson 2003: 72). British soap operas such as *Coronation Street* (1960-present) and *Eastenders* (1985-present) still do not use musical score and have a direct lineage back to the film and television social realism of the early 1960s. It is important not to transplant this tradition across the Atlantic however, given US soap operas are more typically underscored in key dramatic moments.

Hobson also sees a lack of music as 'something which signifies an absence of one of the major elements of the grammar of television' (Hobson 2003: 72). Take away music and the audience may think

1. The hierarchical pre-
ference for dialogue and
music is not just defined
by the sonic prioritisation
within the diegesis but is
also re-enforced by the
division of labour. On
large scale productions, if
there are two re-record-
ing mixers working, the
more senior mixer will
handle the dialogue and
music with the more
junior mixer in charge of
sound effects and atmo-
spheres (backgrounds) as
these are seen as less
important to the finished
product and requiring less
skill to manipulate. Andy
Kris was the sole re-
recording mixer on all
episodes of The Wire.

something is missing. Which brings us to the question of genre. Is *The Wire* a police series or a piece of social realism? Certainly the sound aesthetic is closer to social realism, not being 'the type of police series in which we find action centering on car chases and the sound of racing engines and screeching tyres' (Selby and Cowdery 1995: 66), but this is perhaps where Simon's desire to fill the space the music vacated comes from: he sees the show's strength as its 'realism' and despite it containing some elements of the police series genre, wants to avoid musical score to underline its realist credentials. 'Swear to God, it was never a cop show.' (Simon 2009: 1). If we were to consider *The Wire* as a "cop show", its predecessors such as *Homicide: Life on the Street* (1993–1999), *NYPD Blue* (1993–2005) and *Hill Street Blues* (1981–1987) all used varying amounts of underscore. It is worthy of note that *Homicide: Life on the Street* was most sparing in its use of underscore, a show which David Simon wrote and shares *The Wire's* Baltimore location. The decision to not use underscore in *The Wire* could be seen as a logical extension of the sound aesthetic in *Homicide: Life on the Street*, moving incrementally further from the conventions of the US police series towards social realism.

CINEMATIC PRECEDENTS

Although *The Wire* does use foreground music, it is only in defined musical locations and occasions: the rousing Irish American bar room music at a wake, the gospel church, the dance music in Orlando's strip club. The space vacated by music has elevated background sounds from their status as the least important category of sounds (below dialogue and music) within the aesthetic priority and industrial method of production[1]. There is a clear indication that *The Wire* should not sound like other TV shows.

> I think from meeting one it was clear that they [the creators of *The Wire*] didn't want *The Wire* to sound like a TV show. They wanted a level of detail and attention that was more cinematic, but had a realistic aesthetic.
>
> (Ralston 2009)

This ties in with an overriding trend in cinema sound practice. Steven Handzo states: 'If there is any single trend that can be observed in sound effects practice, it is toward a much more detailed background sound ambience ... On the whole, sound effects have gained at the expense of underscoring' (Handzo 1982: 408). Handzo also identifies Robert's Wise's *Executive Suite* (1954) as perhaps the first Hollywood film to use no score whatsoever, not even during the opening and closing credits. *Executive Suite* does use a more full background than typical productions of the day, emphasising a sense of location but makes no significant linkage between sound and emotion. It is interesting to compare this to Alfred Hitchcock's notes on the use of sound effects in *The Birds* (1963) cited in Sullivan (2006: 262). There is no musical score as such in *The Birds*. Hitchcock gives notes on timbre, rhythm and volume for the use of bird sound effects which could easily be defined as a score, or musical direction, and bear direct comparison to the indeterminate musical instructions of John Cage. Hitchcock's notes would surely have been little different were he directing his composer's music, except they contain the phrase 'we are not using any music at all' (Sullivan 2006: 262). This approach acknowledges the director's desire to remain in control of the audience's emotions during the bird attacks, closing off the audience's room for interpretation. *The Birds* also fills the temporal and frequency domains of the soundtrack with sound effects, much like a traditional score. *The Wire,* by contrast, is sparing in its use of music or sound effects to influence directly the mood of a scene. *The Wire* shares this reliance on sound effects, diegetic music and ambient noise, but this level of formal rigour remains an uncommon approach for screen drama with notable exceptions in the films of Michael Haneke, Ingmar Bergman, Peter Strickland and Paul Thomas Anderson. Jean-Luc Godard's *Vivre sa vie* (1962) is often cited as a key work in the progression towards the use of a more detailed, direct soundtrack, though in direct opposition to classical Hollywood there is a deliberate foregrounding of the apparatus and medium of filmmaking. This is evident in audible changes in the directionality of the microphone during shots and sound cuts which have not been smoothed. We hear the process as much as the sound itself in *Vivre sa vie*; *The Wire* seeks verisimilitude by making the medium as transparent as possible, very much in the Hollywood tradition.

The fact that music can more obviously tell a viewer 'how to feel' does not mean that *The Wire* never uses ambient sounds to do the same job. An example of this is in Season 1 Episode 10 that contains two scenes with Bubbles which bookend the episode. Both scenes feature Bubbles sitting on the same park bench watching the world go by – or more accurately *listening* to the world go by. In the first scene, he sits in bright sunshine and from the previous episode we are aware he has just come off drugs. He sits, unsure of himself, listening to the sounds of children playing, birds in the trees and gentle breezes as though they were alien to him. The scene suggests his re-awakening to sensory experiences excluded by the effects of drugs; and also with its deliberate use of 'innocent' sounds, suggests him returning to his own childhood to a time before his life was tainted by drug abuse. This is communicated without music – the sound effects have become the emotional signifiers within the soundtrack, as well as giving us a 'real ambient sense of the moment' (Simon, 2004).

However, at the end of the episode the scene is reprised after dark. The innocence has been replaced by the temptation of drugs as the children's voices have been replaced by the shouts of drug touts and the birds replaced by distant sirens. In both scenes there is neither prioritised dialogue nor music, and the background atmosphere has been composed for its ability to support the prevailing emotional tone of the character at that point in the narrative. David Simon may suggest that *The Wire* does not use score: I suggest it does on occasion, but it is a score where the instruments are sound effects, snatches of dialogue and ambient backgrounds; the composers of this score are the post-production sound team. This crossover into a form of *musique concrète* disguised as sonic

2. See 'Multi-Channel Dialo-
gue and Effects Recording
During Film Production'
by James E. Webb Jr. in
American Cinematographer,
April 1979.

3. The detail of the technical
set up used by Bruce
Litecky on *The Wire* is
described on his website,
www.locationsoundservi-
ces.com accessed 5 Feb-
raury 2009.

verisimilitude is not unique to *The Wire*, but it has typically been a cine-matic rather than televisual device.

THE WIRE'S SOUND PRODUCTION

In the early years of synchronous sound and moving image, cameras were limited to studio use because of size and weight, and 'clean' sound was only achievable in the sonically controllable soundstage. When cameras sub-sequently became portable enough for location use, the technology of location sound was initially unable to deliver an appropriate result; location sound was often replaced in post-production. Portable, high quality tape recorders such as the Nagra, which emerged in the late 1950s, allowed sound recording a degree of portability and quality it had not previously enjoyed, but retained the problem that locations themselves were fre-quently too noisy. With the advent of Dolby noise reduction, radio microphone systems and analogue multi-track recording – which can isolate individual characters' dialogue for flexible mixing (pioneered by Robert Altman and Jim Webb on *California Split*, (Altman 1974)) – sound and image could be recorded concurrently in most locations with sufficient separation from ambient noise as to be useable in a fictional drama[2]. Webb used a system the size of a hospital trolley requiring several assistants, which was only affordable by well-financed feature film pro-ductions because of the cost of the equipment and the extra personnel. Nearly forty years later, equivalent capabilities are provided by digital multi-track and modern radio microphones that are affordable to almost any production. Their size permits portable use when required and recording quality is unprecedented. (Although that often means pristine reproduction of sounds you didn't want on the soundtrack.) In the case of *The Wire*, Bruce Litecky, the Production Sound Mixer used a multi-track digital recorder and six radio microphones as well as the boom micro-phone[3]. However, the shooting style determined how much of the boom microphone could be used. 'Because most of the scenes were shot with two cameras, blocking and boom placement was extremely difficult. Which is why the majority of the dialog was recorded with radio mics.' (Kris 2009)

Radio microphones provide better separation between actors and iso-lation from background noise than can be achieved with boom micro-phones. But it requires a great deal of post-production manipulation to achieve an audio aesthetic which makes its technical apparatus appear largely transparent, what Mary Ann Doane calls 'reducing the distance perceived between the object and its representation' (Doane 1982: 164). While this distance can be reduced, it cannot be eliminated: by definition this is an industry built on representation, not actual experiences. The use of radio microphones initially makes the representation further from the object, and only with further post-production work can the distance be reduced. The separation given by radio microphones gives an unnaturally close sound perspective (imagine the sound if your head was inside someone's jacket, where microphones are typically hidden). This per-spective is corrected at the sound editing and mixing stages. Footsteps and ambient noises are artificially low in relation to dialogue and must therefore be added by a Foley artist or from a sound effects library. And the goal is always to make the process as transparent as possible, even if it means adding additional noise.

> You go through a scene, find the worst-sounding line of dialog in it and then go through and make the rest of the scene sound as bad as that. Because a consistently bad-sounding scene is less distracting to the ear than a scene that goes from crystal clear to suddenly noisy and back.
>
> (Ralston 2009)

The freedom afforded modern productions by the small size and cost of radio microphones and multi-track digital recorders increases the amount of post-production sound work required to remove the audible signatures caused by their use ... This is quite a different sound aesthetic to that championed by the Dogme school, whose rules included not allowing post-production manipulation of the soundtrack: whether this type of intervention is less truthful as implied the Dogme, the 'Vow of Chastity' is open to debate. The second statement in the Dogme 'Vow of Chastity' is:

> 2. The sound must never be produced apart from the images or vice versa. (Music must not be used unless it occurs where the scene is being shot).
>
> (Von Trier and Vinterberg 1995)

Lars von Trier's *The Idiots* uses on set mixing and positioning of directional microphones in order to produce a soundtrack which might be described as 'controlled verité'. It privileges the location sound to a higher degree by removing the safety net of post-production work. The integrity of location sound and image are equally vital to the story telling. If *The Idiots* has given the *recording* of sound concurrently with image a higher priority, it has largely reduced the soundtrack to being in a simplistic cause and effect relationship with the image. *The Idiots* moves the point of control and the toolset from post-production to the set. In doing so, it highlights the 'recordedness', the material fact of the sound's relationship to the device that captures it. *The Wire* also attempts to represent truth with its soundtrack but places control more in post-production. This does not intrinsically lessen *The Wire's* ideals, merely the point of control is different. If a noise is eradicated by removing its source from the set (as in *The Idiots*) this is surely a different *type* of intervention than adding a footstep or filtering a hum in post-production, not a more or less 'truthful' one (*The Wire*).

THE IMPORTANCE OF DELIVERY

There have been major changes in how consumers receive, view and hear television over the past twenty years, with increased quality in both

delivery formats and typical consumer's viewing and listening technology.
For many film and TV creators, the issue of how your final soundtrack
translates to a variety of playback systems has been a problematic one.
The final sound mix, technically operated by the re-recording mixer but
attended by other members of the sound team and in *The Wire's* case,
the show's executive creative team, is perhaps where it sounds as good
as it ever will. Taking place in a facility designed to offer an idealised
listening environment, typically the size of a small cinema, a TV show
played on domestic equipment will typically sound worse, or at the very
least different. A TV re-recording mixer might also make sure the finished
show translated well to the small, poor quality monophonic television
speakers that used to be the norm for home listening. However, *The Wire*
was mixed and after Season 1, aired in the Dolby Digital 5.1 surround
format, originally used for theatrical presentation of feature films and many
viewers have also found DVD box sets as a new way to enjoy television
shows. This means that the delivery of the TV soundtrack is shifting more
towards cinema technology, at least for a significant minority of the
audience who have home cinema equipment. The soundtrack is also much
closer to what was heard during the final sound mix if played through such
a system. The Dolby Digital format offers both a better translation from
mixing theatre to viewer as well as the use of additional sound channels for
the home listener.

> For me the best part about mixing and delivering in 5.1 was that the
> viewer heard the mix as we mixed it, without going through all the
> crazy broadcast compressors and limiters that most analog broadcasts
> go through.
>
>   (Kris 2009)

These 'crazy broadcast compressors and limiters' reduce dynamic range
and drastically affect the way a mix is heard, effectively turning down the
loud parts and turning up the quiet parts, upsetting the delicate balance
created during the sound mix. Freed from this (necessary) limitation of
analogue transmission, TV productions have the ability to produce sound
mixes with a wider dynamic range closer to a cinematic experience. I would
suggest *The Wire's* sound aesthetic has partly evolved as a result of the
format it can be delivered in, either through digital broadcasting or DVD
box sets. The basic approach of 5.1 mixing in the show is to provide a
series of direct events (on-screen or off-screen but part of the diegesis) at
the front speakers with additional ambience, reverberation and occasional
far off-screen dialogue in the rear speakers. Surround sound is not always
about helicopters and bullets flying around your ears in all directions, it can
simply function as an aid to 'immersion' through a wide soundfield of
steady-state ambiences.

> ... in 5.1-channel sound, reverberation is likely to appear in all
> five channels. After all, good reverberation is diffuse, and it has
> been shown that spatially diffuse reflections and reverberation

contribute to a sense of immersion in a sound field, a very desirable property.

(Holman 2000: 31)

The increased possibilities of surround (at least for those with capable systems) mean that the TV soundtrack can use a form which has previously only been considered viable in a theatrical setting. The crossover of the show's aesthetics, production methods and presentation blurs the boundary between the cinematic and the televisual.

WHERE DID THAT COME FROM?

The use of mono, stereo or surround soundtracks bring with them attendant differences in how images can be perceived by a viewer/listener. When one listens to a mono soundtrack, the focus of your attention is purely on sound emerging from the front – assuming the speakers are behind or close to the image. Hence your perception will strive to reduce and ignore other sounds from the sides and rear. This does not mean that sounds coming from the screen are only recorded in one direction, but they are replayed from one direction. Mono films can condense 360° of sound and replay it from a single speaker which is in one sense more mediated than surround sound: sound is stripped of its directionality A footstep moving behind the camera position becomes quieter and more reverberant (interpreted as distant), but remains emerging from the same direction (the screen) rather than behind the audience. This is a defining characteristic of the monaural track, its inevitable condensing of 360° of sound into a point source replayed behind or directly adjacent to the screen.

In the surround sound experience, by positioning sounds from the rear and sides, the creators are asking the audience to be receptive to sounds existing physically outside of the screen. This divergence of sound radiation will inevitably create an additional dimension for sound not possible with the image – sound and image are no longer constantly radiating from the same direction. Sound related to the image now radiates from a greater number of directions; this has the potential to be a closer match between the recorded replay system and the actual event

> When we are on the streets of Baltimore at night and you hear people shouting from buildings, drug dealers selling their wares, dogs barking, distant sirens, car alarms, shootings from two blocks away, kids playing ... it's so real and so terrifying that it makes the viewer realize in a very unobtrusive (but if you listen closely, very obtrusive) way that things in Baltimore are 'messed' up big time.
>
> (Kris 2009)

The drive by sound of in-car entertainment systems is something that relies on the increasing spatial possibilities of surround sound as well as the subwoofer speaker. In Season 4, Marlo Standfield's SUV is identified by its booming bass, with Bodie and other corner boys reacting to its approach.

The ability of a capable home cinema system to provide the deep bass notes needed to suggest the in-car entertainment system acts as an interesting linkage between character and the viewer/listener. The subwoofer is found in car entertainment and also home cinema: without the home cinema ability to resolve these frequencies, the effect is diminished for the viewer/listener because, in a sense, the characters have better sound systems than they do. In the context of a more action driven show, the use of frequent loud and bass heavy effects can lessen cumulatively their impact, but in *The Wire* the usual scarcity of deep low frequencies prime the audience to be alert to a character's arrival into a scene or as a brief distraction as they fly by from side to side, with reverberant characteristics added to match. Jen Ralston:

> Andy Kris chose the Altiverb plug-in to create the spatialization effects for the music cues. If I recall correctly, he came across this plug-in while we were working on Season 3 and was very pleased because until then it was a time-consuming process for him to get the music to sit in each location. There was no set method before Altiverb came on the scene.
>
> There are a few drive-by stereos that our music editor, Blake Leyh (a genius sound designer in his own right) recorded by literally putting the music in a car and having it pass by. Since clearances for music were usually the last things to get settled on in the mixing process, Blake's drive-bys were all his own personal compositions and generally only about 30 seconds long. I don't believe we began employing that technique until Season 4 though. Up until then, it was all Andy on the mix stage.
>
> (Ralston 2009)

The approach to achieving a realistic acoustic imprint for these drive-bys shows pragmatism towards achieving the required image/sound match, using a mixture of software and physical world sound design techniques.

The fact that content creators are working in a surround format does not prevent single point sounds, since it is also possible for the mono track to exist inside of the stereo or 5.1 surround track and be used as a stylistic device in support of the storytelling.

> The biggest discussion I can recall about the way to approach sound stylistically was with the surveillance photography: Normal state of things - the show is in color, the sound is 5.1. We go into the POV of the telephoto surveillance SLR camera and the perspective of the sound pulls back in distance to match it and folds in a little to mimic the tunnel vision effect. The camera CLICKS off a photo and the action freezes in black and white for half a second, all the specific sounds drop off and only a mono air tone remains. That was the road map we made in Season 1 for the photos.
>
> (Ralston 2009)

Here the use of a mono 'moment' has allowed the suspension of time – the atmosphere present in all of the speakers is removed because time has stopped for the moment of snapshot. Localised sound can also be used as a method analogous to the visual zoom. Scenes which begin with a wide shot typically feature a wide distribution of ambient sounds in all speakers. On cutting to close-up dialogue between the characters, the balance shifts and dialogue is privileged. This ability to 'zoom' sound is perfectly possible with stereo, but with 5.1 surround it creates a perceptual shift in terms of direction. We hear five speakers and then the focus is shifted to the front located sound as the sound mix dictates.

THE AUTHORED SOUND WORLD

It is important to consider how the authorship of *The Wire's* soundtrack is arrived at. On the one hand we have a clearly defined creative figurehead, David Simon, in his multiple roles as writer, creator and executive producer, on the other we have a variety of people involved in the production who also contribute as 'authors' of events within the show.

We need to define Simon's concurrent roles as separate entities, since each role influences the show in different ways. As the creator, his role determines the overall aesthetic and consistency between episodes and seasons, such as making clear at the start that there would be little or no use of non-diegetic music. As a writer, Simon would impart meaning into the show's soundtrack using sound as a storytelling device. In this instance, authorship extends from the idea right through to the finished show. But without Simon then taking on the role of Executive Producer, maintaining creative direction through shooting and into post-production, there is obvious potential for his original intent to be lost. This level of control establishes a notion of an over-arching single author structure at the top of the industrial hierarchies of the production team. Simon's creatorial authorship is maintained by his insistence on continuity of approach to visual and sound style based on aesthetic rules which were established from Episode 1 Season 1.

However, despite this semblance of a creator responsible for everything in *The Wire*, the show has multiple writers, directors and editors and a large crew. It would be wrong to ignore the influence of collaboration and *The Wire's* soundtrack is perhaps more collectively authored than other aspects of the show.

> This being television, the one crew member who was always absent [from the final sound mix] was the director of the show, which I thought was odd, but after mixing a few episodes, I realized that it was the executive producers who were really in charge of the finished product. They were the ones who had the most input and knew the arc of the entire season, so they could comment on something that would affect a future episode in a way that the director (who is usually different from show to show) couldn't.
>
> (Kris 2009)

4. The use of non-actors for medical scenes is covered in 'Bullitt: Steve McQueen's Commitment to Reality' (1968) available on *Bullitt* (Two Disc Special Edition) 2005, Warner Home Video.

Simon is not a sole author even at the sound mixing stage, with the other executive producers contributing their opinions. Nor did a complete set of rules emerge which could be applied as a template to the show's sound aesthetic.

> We didn't have any hard and fast rules for doing the things we did. We would try things different ways (like more reverb, less backgrounds, etc.) until everyone felt comfortable with where we got. It was always trial and error and we knew we got it when it just felt "right".
>
> (Kris 2009)

At this level we are still talking about a small group of professional people arriving at the collaborative finished product, each from a defined hierarchical position in the decision-making process. As Sellors (2007) notes:

> Some sound recordists will count as authors under a notion of collective filmic authorship while others will not. It will depend on the recordist's contribution to the filmic utterance ... In order to know whether the sound recordist counts among a film's collective authors, we need to understand this person's role in producing not just the material film, but also its utterance. We must equally be careful to consider the consequences the complex nature of film has for the planes of authorship that can occur in film production.
>
> (Sellors 2007: 269)

The importance of collective authorship is not the hierarchy, but whether an individual produces what Sellors calls a 'filmic utterance', taking his definition from Grice (1969), who defines an utterance as an 'intentional (purposeful), meaningful expression'. (cited in Sellors 2007: 270)

These utterances could potentially come from the notion of a group of 'insiders', individuals with specialised knowledge and experience of the subject matter who are employed to add authenticity to events within a piece of fiction, shifting the centres of authorship in the process. Peter Yates's *Bullitt* (1968) contains an example of this. Frank Bullitt watches through the observation panel of an operating theatre as his injured colleague, Ross, is operated on. In this operating theatre practising surgeons and nurses were used in lieu of actors in pursuit of authenticity[4]. The sound for this scene is formally quite different from the rest of the film using a pseudo-documentary observational mode. Dialogue flows from all of the participants and overlaps. The noise of clattering medical instruments obscures words. No attempt is made to 'project' the voice and much of what is said is either inaudible or specialised medical jargon. On completing the operation, one of the surgeons emerges and the film then returns to a classical Hollywood style of dialogue as Bullitt and the lead surgeon discuss the patient's condition, each taking turns to speak clearly and without overlap.

The fact that the entire operation scene is from Bullitt's point-of-view helps motivate its shift in tone, but it remains a documentary episode contained within a dramatic film. It highlights that the film's soundtrack follows the classical Hollywood model but with excursions into documentary technique and aesthetic. *The Wire* also uses the 'insider' to add authenticity, but in a different context. Unlike Bullitt, *The Wire*'s insider contributions are more layered and interwoven with the narrative, not compartmentalised. As Jen Ralston relates:

> We recorded an enormous amount of background walla [background voices] for the show because the commercial library wallas don't really cover the demographics or feel of the standard homicide office, blues club, West Baltimore drug corner or Stevedore's bar.
> The 'Wire loop group' was the real deal. They understood the world of *The Wire*. They live in Baltimore. They have an ear for the accents and speech patterns. We had several seasoned voice actors as well as a group I'll refer to as 'The Corner group' because they were some of the real people that Executive Producers David Simon and Ed Burn's book-turned-HBO mini-series 'The Corner' was based on.
> The first session I worked with the 'Corner group' was mid-way through the first season. Up until then, the recording sessions were very structured: a line was cued, a line was written, a line was recorded after the three beeps. But it was all sounding very lifeless to our ears at the mix. So I went down to Baltimore and started working with them. They had a list of suggested lines for a prison scene: 'Hey man, get off of me,' 'Yo, guard, I need to go to the infirmary,' and the like.
> Not being experienced actors, they tended to sound like they were reading when they referred to the script page. So I took the script away and told them what I wanted them to say and they mimicked my delivery. That went well, but they were all kinda smirking as they did it. I asked what was up and they explained that just wasn't how they would say it. So I told them to say it their way. 'Guard' became 'Turn-key,' 'get off of me' became 'back the fuck off me.'
> In matters of the language of the street, the writers deferred to the loop group during those sessions. David would suggest names for drug products but we always recorded many more names than he scripted based on what the group was actually hearing being touted on the streets that week.
>
> (Ralston 2009)

From this we see that every background sound had to have the authenticity of being recorded in Baltimore, and that the process of arriving at convincing voices for background required an inverted authorship model. Lines were written, passed to Jen Ralston for recording, then the performers revised them according to their own ideas of authenticity. The Corner Group have now become cinematic authors, with

Ralston acting as an intermediary and performance director in between them and David Simon. Although it might be considered that the background sounds of *The Wire* are of much less important than other aspects of the show (lead actors performances, visual style, writing) I would suggest that with the lack of score and the central role of verisimilitude in the piece, they are a bedrock of authenticity because of their elevated status in the diegesis. It is interesting to note that the insider's authorship can be quite removed from the show's creator, even if he remains in overall executive control.

The attempt by the filmmaker to create a soundtrack that can be readily accepted as a reality is as old as synchronous sound, but the pursuit of that goal is a constant challenge for different generations. *The Wire*'s soundscape is both singly and collaboratively authored, individual and industrialised. It is progressive in terms of presenting a form of sonic verisimilitude that avoids the self-reflexive nature of some conceptually driven soundtracks. The various ways in which the show has used technology, aesthetic rules and collective authorship to achieve its overriding goal of sonic verisimilitude does not allow easy categorisation or attachment to any one prior school of thought or methodology. It is not a rigid set of commandments as in the Dogme approach, nor an adherence to a mechanical process (or subversion of process) as typified by Godard. *The Wire* achieves a sophisticated and believable sound world because of and in spite of recent technologies; it sets out clear rules about the use of music and then breaks them when it suits the format of the show. But by remaining pragmatic and relying on a degree of instinct and professional judgement, *The Wire* is on its own terms a very successful demonstration of how verisimilitude can be achieved, conveyed and employed in television sound, often repurposing elements from the cinematic.

SOURCES

Altman, Robert (1974), film, *California Split*, USA: Colombia Pictures.

Attanasio, Paul (1993–99), TV series, *Homicide: Life on the Street*, USA: NBC.

Bainbridge, Caroline (2007), *The Cinema of Lars von Trier: Authenticity and Artifice*, London: Wallflower Press.

Bjorkman, Stig (ed.) (2003), *Trier on von Trier*, London: Faber and Faber.

Bochco, Steven and Michael Kozoll, (1981–87), TV series, *Hill Street Blues*, USA: NBC.

Bochco, Steven and David Milch (1993–2005), TV series, *NYPD Blue*, USA: 20th entury Fox Television.

Collet, Jean (1972), 'An Audacious Experiment: The Soundtrack of *Vivre sa vie*' in Brown, Royal (ed.), *Focus on Godard*, New Jersey: Prentice Hall, pp. 160–162.

Doane, Mary Anne (1985), 'The Voice in the Cinema: The Articulation of Body and Space' in Weiss, E. and Belton, J. (eds) *Film Sound: Theory and Practice*, New York: Columbia University Press, pp. 162–175.

Godard, Jean Luc (1962), film, *Vivre sa vie*, France: Les Films de la Pléiade.

Handzo, Steven (1982), 'Glossary of Film Sound' in Weiss, E. and Belton, J. (eds) *Film Sound: Theory and Practice*, New York: Columbia University Press, p. 408.

Hitchcock, Alfred (1963), film, *The Birds*, USA: Universal Pictures.

Hobson, Dorothy (2003), *Soap Opera*, Cambridge: Polity Press.

Holman, Tomlinson (2000), *5.1 Surround Sound: Up and Running*, Boston: Focal Press.

Kris, Andy (2009), Email Interview with the author, February 2009.

Ralston, Jen (2009), Email Interview with the author, February 2009.

Schreger, Charles (1982), *Altman, Dolby, and the Second Sound Revolution* in Weiss, E. and Belton, J. (eds), *Film Sound:Theory and Practice*, New York: Columbia University Press, pp. 348–355.

Selby, Keith and Ron Cowdery (1995), *How to Study Television*, Macmillan: London.

Sellors, C. Paul (2007), 'Collective Authorship in Film', *The Journal of Aesthetics and Art Criticism*, 65: 263–271. doi: 10.1111/j.1540–594X.2007.00257.x

Simon, David (2004), DVD commentary track from Episode 1 of *The Wire*, USA: HBO Video.

Simon, David (2009), 'Introduction', in Alvarez, Rafael (2009), *The Wire Truth be Told*, Edinburgh: Canongate.

Smith, Julia and Tony Holland (1985-present), TV Series, *Eastenders*, UK: BBC TV.

von Trier, Lars (1998), film, *The Idiots*, Denmark: Zentropa Entertainments.

von Trier, Lars and Thomas Vinterberg (1995), *Vow of Chastity*, http://www.dogme95.dk — accessed 10 February 2009.

Warren, Tony (1960-present), TV series, *Coronation Street*, UK: Granada Television.

Wise, Robert (1954), film, *Executive Suite*, USA: Warner Brothers.

Yates, Peter (1968), film, *Bullitt*, USA: Warner Brothers.

CONTRIBUTOR'S DETAILS

Robert Walker is a sound designer, re-recording mixer and composer with credits in feature films, television, short films, animation and installations. He currently teaches Film Sound at Screen Academy Scotland in Edinburgh and is a member of the Association of Motion Picture Sound. Recent publications include a chapter in *Mark E. Smith and The Fall: Art, Music and Politics* (Ashgate, 2010) and 'Cinematic Tinnitus' in *Reverberations* (Continuum, 2012). He is also completing his first feature film as a writer/director, entitled 'RSJ'.

Contact: Screen Academy Scotland, 2a Merchiston Avenue, Edinburgh, EH10 4NU
rj.walker@napier.ac.uk

PHILIPPA LOVATT

'Every drop of my blood sings our song. There, can you hear it?': Haptic sound and embodied memory in the films of Apichatpong Weerasethakul

ABSTRACT

Frequently drawing on avant-garde formal strategies, bringing together personal, social and cultural memories in a cinematic collage, the films of Thai director Apichatpong Weerasethakul recreate what Richard Dyer has called 'the texture of memory' (Dyer 2010). Using narrative techniques such as repetition, fragmentation, and convergence (as different threads of a narrative resonate uncannily both within and across the films), the work expresses what the process of remembering *feels like, how the warp and weft of the past continuously moves through and shapes the present just as the present shapes our memories of the past. While sound design in classical cinema often privileges the voice, lowering ambient sound in order to ensure intelligibility while creating an illusion of naturalism, in these films 'natural' ambient or environmental sounds are amplified to the extent that they become almost* denaturalized, *thus*

The New Soundtrack 3.1 (2013): 61–79
DOI: 10.3366/sound.2013.0036
© Edinburgh University Press
www.euppublishing.com/SOUND

KEYWORDS

sound
haptic
phenomenology
affect
memory
Thai cinema
Apichatpong
    Weerasethakul
spectatorship
embodiment

heightening their affective power. In Blissfully Yours *(Sud sanaeha, 2002),* Tropical Malady *(Sud pralad, 2004),* Syndromes and a Century *(Sang sattawat, 2006) and* Uncle Boonmee Who Can Recall His Past Lives *(Loong Boonmee raleuk chaat, 2010) the sound of the environment is often so dominant that it dismantles our reliance on the verbal or the linguistic to ground our understanding of what is happening in the narrative, and instead encourages (or rather* insists upon*) an embodied, phenomenological, engagement with the sensuality of the scene. This use of sound and textual synaesthesia foregrounds sound's materialism and its relationship to touch, sight, and taste, creating a feeling of sensory immersion on the part of the spectator where the senses seem to become indistinct. Alongside frequent bursts of pop music (expressing* jouissance*), the films' sound designer Akritchalerm Kalayanamitr uses these environmental sounds to create rhythmic 'sonic sequences' that have themselves an almost musical quality reminiscent of experimental avant-garde compositions from the 1950s and 60s made up of single or multi-tracked field recordings. This essay examines these moments in Apichatpong's films and argues that they enable a sense of connection and intersubjectivity by appealing directly to the audio-viewer's shared knowledge of how we remember.*

While sound design in classical cinema often privileges the voice, lowering ambient sound in order to ensure intelligibility, creating an illusion of naturalism, in the work of Thai artist and filmmaker Apichatpong Weerasethakul, 'natural' ambient or environmental sounds are amplified to the extent that they become almost *denaturalized*, thus heightening their affective power. In the feature films *Blissfully Yours* (*Sud sanaeha*, 2002), *Tropical Malady* (*Sud pralad*, 2004), *Syndromes and a Century* (*Sang sattawat*, 2006), *Uncle Boonmee Who Can Recall His Past Lives* (*Loong Boonmee raleuk chaat*, 2010), and the short films *Mobile Men* (2008), *Ashes* (2012b) and *Cactus River* (*Khong Lang Nam*, 2012), the sound of the environment is often so dominant that it dismantles our reliance on the verbal or the linguistic to ground our understanding of what is happening in the narrative, and instead encourages (or rather *insists* upon) an embodied, phenomenological, engagement with the scene. Recognising the permeability of the imaginary line between the spectator's body and the 'body' of the film forms the basis of important recent developments in film studies and theories of spectatorship (Barker 2009, Marks 2000, Sobchack 2004). This article builds on some of these ideas and asserts that a focus on the sonic can significantly enrich our understanding of the cinematic experience. In contrast to earlier 'visually orientated' models of spectatorship then, I argue that an exploration of the intersubjective and affective properties of sound opens up the possibility of an ethical spectatorship based on listening.

Apichatpong's films and video work present a rich tapestry of storytelling traditions: folklore, *likay* folk theatre, soap opera, horror movies, adventure stories and science fiction − all of which have, and continue to play, apart in the formation of the Thai cultural imaginary. These frames of reference are broadened further as his work also demonstrates a number of Western influences including American structural and avant-garde filmmaking and European art cinema (Ingawanij and MacDonald 2006). Born in Bangkok in 1970, Apichatpong grew up in the town of Khon Kaen in Northeast Thailand and studied architecture at Khon Kaen University before completing a Masters of Fine Arts at the School of Art Institute in Chicago where he made his first short films in 1994. On returning to Bangkok, he formed the independent production company, Kick the Machine, and made his first feature film, *Mysterious Object at Noon* (*Dogfar nai meu marn*, 2000). Since then, he has gone on to make several feature films, shorts and video installations exhibiting his work both nationally and internationally. He has been an active supporter of Thailand's independent film culture, co-directing the fifth Bangkok Experimental Film Festival in 2008. His most recent full-length feature film, *Uncle Boonmee Who Can Recall His Past Lives*, won the coveted Palme d'Or at Cannes in 2010.

While Apichatpong's work has been the subject of a great deal of critical attention, particularly following the success of *Uncle Boonmee Who Can Recall His Past Lives*, the importance of sound in his films and his collaborations with sonic artists Akritchalerm Kalayanamitr and Koichi Shimizu have rarely been discussed.[1] Akritchalerm has been the sound designer on all of Apichatpong's films since *Tropical Malady*. Born in Bangkok in 1975, he studied Political Science and International Affairs at Thammasat University, before going to study filmmaking in San Francisco where he graduated from film school in 2000. Since returning to Thailand, Akritchalerm has worked on a number of important films including Pen-ek Ratanaruang's *Ploy* (2007) and *Nymph* (2009), Aditya Assarat's *Wonderful Town* (2007), Anocha Suwichakornpong's *Mundane History (Jao nok krajok*, 2009) and Naomi Kawase's *Nanayomachi* (2008). He collaborated with Koichi Shimizu on the sound and video installation *Anat(t)a* (2006–8), which was exhibited in Bangkok and Rotterdam at the 37th International Film Festival. Born in Japan in 1972, Shimizu studied audio engineering in New York from 1991 to 1993 before moving to Bangkok in 2003 where he has worked as a music producer, multimedia artist, and composer for television commercials and films including Pen-ek Ratanaruang's *Invisible Waves* (2006) and *Ploy* (2007), and Aditya Assarat's *Wonderful Town*. Credited alongside Akritchalerm as a sound designer, he also composed the scores for Apichatpong's *Syndromes and a Century* and *Uncle Boonmee Who Can Recall His Past Lives*.

Bringing together influences and sources from a wide variety of storytelling traditions, Apichatpong's films blur the boundaries of personal and social memory. Using narrative techniques such as repetition, fragmentation and convergence (as different threads of a narrative resonate uncannily both within and across the films), the work expresses what the *process of remembering* feels like, how the warp and weft of the past continuously moves through and shapes the present just as the present shapes our memories of the past. Made during a period of continued political unrest in Thailand, in a culture policed by strict *lèse majesté* censorship laws, the films tend to focus on the experiences and memories of those on

1. A notable exception is May Adadol Ingawanij's conference paper 'Sounds from life and the redemption of experience in Apichatpong Weerasethakul's films', delivered at the *Screen* Conference, Glasgow, 2008.

2. See David Teh's 'Itinerant
   Cinema: The Social Surrealism of Apichatpong
   Weerasethakul', *Third
   Text*, Vol. 25, Issue 5,
   2011, 595– 609, for a
   detailed account of the
   historical and political
   context that informs Apichatpong's work

the social and political periphery, such as characters from Thailand's impoverished northeast, Burmese migrant workers, gay men, older women and children, whose voices are generally absent from public discourse. His more recent work, the *Primitive Project*, moves towards a more explicitly historical framework as it is concerned with memories of Thailand's violent past that have largely been 'forgotten' in official records[2].

In film studies, the relationship between memory and representation is most often described in visual terms (the use of certain editing techniques such as the flashback, for example) and yet, the focus on the cinematic image misses the mnemonic potential of the sonic. Like sand disappearing through the hourglass, sound cannot be held still. As Walter Ong describes, 'Sound exists only when it is going out of existence. It is not simply perishable but essentially evanescent, and it is sensed as evanescent' (Ong 2002: 32). The evanescence of sound makes it a rich metaphor through which to explore the transient, and often, involuntary nature of memory. However, the sonic realm also has 'concrete', material properties that affect both the body and the imagination of the listener. As Walter Benjamin describes in 'A Berlin Chronicle':

> The déjà vu effect has often been described. But I wonder whether the term is actually well chosen, and whether the metaphor appropriate to the process would not be far better taken from the realm of acoustics. One ought to speak of events that reach us like an echo awakened by a call, a sound that seems to have been heard somewhere in the darkness of a past life ... [T]he shock with which moments enter consciousness as if already lived usually strikes us in the form of a sound. It is a word, tapping, or a rustling that is endowed with the magic power to transport us into the cool tomb of long ago, from the vault of which the present seems to return only as an echo.
>
> (Benjamin 2007: 59)

Benjamin's words make a powerful connection between memory's flow and the affective properties of particular sounds, which act as mnemonic triggers. Through an analysis of the sound design of Apichatpong's films, this article similarly explores the materiality of sound – in particular, its rhythms, tones and timbres – in an attempt to understand, not what these sounds might 'represent', but how the affective power of the sound design might capture a sense of *how it feels* to remember.

Alongside frequent bursts of exuberant pop music in the films, Akritchalerm uses environmental sounds to create rhythmic 'sonic sequences' that have themselves an almost musical quality reminiscent of experimental avant-garde compositions from the 1950s and 60s made up of single or multi-tracked field recordings such as those by Steve Reich and John Cage. This artistic practice in America was mirrored in Europe by the work of *musiqueconcrète* pioneer Pierre Schaeffer who made compositions using tape recorders and 'found' sounds, and later in Canada with the introduction of the 'World Soundscape Project' led by R. Murray Schafer.

Influenced by Husserl, Schaeffer developed a phenomenological approach to sound analysis that was interested in describing the perceptual qualities of a sound rather than attaching it to a source and the information it might convey. As he explains: '[t]he dissociation of seeing and hearing ... encourages another way of listening: we listen to the sonorous forms, without any aim other than that of hearing them better, in order to be able to describe them through an analysis of the content of our perceptions' (Schaeffer 2005: 78).

Through this technique, Akritchalerm's soundtracks mediate and transpose the soundscapes of lived space into the cinematic experience, communicating a sense of character interiority and perception through the use of subjective sound and point-of-audition. In combination with the image, the composition of electronic scores and 'found' field recordings foregrounds sound's materialism – its 'concreteness' – and its relationship to touch, sight, and taste. This form of textual synaesthesia encourages a feeling of sensory immersion on the part of the spectator as the senses become indistinct. The perceived permeability of the imaginary boundary between the 'body' of the film and that of the spectator is a common response to Apichatpong's audio-visual aesthetic. As Graiwoot Chulphongsathorn writes of his first experience watching *Tropical Malady*: 'After the credits had ended, I wanted to embrace the film and slowly melt into it. Momentarily, I did not exist and felt no different from the wind in the middle of the jungle at night' (Chulphongsathorn 2004). Sound theorist Brandon LaBelle claims that listening makes permeable the boundary between self and other partly because of its close relationship to touch and the way its presence is felt on and through the body; according to the laws of physiology, a sound wave only becomes a sound when it reaches and vibrates the bones in the inner ear (Ashmore 2000: 65). It is perhaps this intimate connection of sound to our bodies, I will suggest, that makes it particularly able to create a sense of commonality and sensory exchange in cinema.

The link made by Graiwoot above, between the sound of a film and the spectator's embodied perception of it, has been theorised by Laura U. Marks as 'haptic hearing' (Marks 2000: 183). By foregrounding the 'texture' of sound using techniques such as excessive amplification, vibration or distortion, sound design can communicate 'feeling' through its close association with the sense of touch, and by extension, emotion (Coulthard 2012). Apichatpong's short film *Mobile Men* (2008) is an example of how haptic sound can be both viscerally and emotionally powerful. Made as part of the 'Art for the World: Stories on Human Rights' project, Apichatpong describes the film as a 'portrait' of Jaii, a migrant worker from Burma. Intended to 'instil and capture his confidence and dignity', the filmmaker gives the hand-held camera over to Jaii and his Thai companion, Nitipong, on the back of a moving truck allowing them to film themselves, and to some extent, take ownership of their own representation (Weerasethakul 2009). As I shall discuss in more detail later, in Apichatpong's work, extended driving sequences like this communicate a feeling of transcendent rapture, a few moments of ecstatic transformation where the subjects of the films are given temporary reprieve from the oppressive realities of everyday life. For Apichatpong, the truck in *Mobile Men* becomes 'a small moving island without frontiers where there is freedom to communicate, to see and to share' (Weerasethakul 2009). As the truck speeds along, the sound of the wind dominates the soundscape as it whips ferociously around the microphone creating an atmosphere of violence and intensifying pressure. Nitipong mutely points to his lips, then to the logo on his chest, the stitching on his jeans and finally down to his trainers where he points at similar details silently gesturing towards the products of a cheap migrant labour force exploited by capitalism. He then stands up, removes his shirt and begins to strike various 'strong-man' poses. Jaii then points to his tattoos (which, he shouts over the wind, were intended to impress girls) and, laughing, tells us that having them done was agony. Jaii rips off the microphone taped to his chest, and attaches it to the tattoo on his upper arm, symbolically connecting voice or perhaps more accurately, *voicelessness*, with pain. He then throws back his head and lets out a gut-wrenching primal roar that is both a release of tension and a desperate cry of protest that is ultimately lost on the wind.

With haptic sound, Marks explains, 'the aural boundaries between body and world may feel indistinct: the rustle of trees may mingle with the sound of my breathing, or conversely the booming music may inhabit my chest cavity and move my body from the inside' (Marks 2000: 183). In the analysis that follows, I attempt to develop these ideas as I demonstrate how the soundscapes of Apichatpong's films enable a sense of connection and intersubjectivity by appealing directly to the spectator's embodied self. My approach also engages with Felicity Callard and Constantina Papoulis' claim that theories of affect move discussions of memory on from 'an understanding of subjectivity and of experience that is based on an internal world, on particular formulations of memory and representation' towards a concern with the 'nonrepresentational and extralinguisitic aspects of subjective experience' (Callard and Papoulis 2010: 247). Importantly therefore, I will demonstrate how haptic sound also allows us to move away from questions of signification towards a closer understanding of our embodied engagement with the acoustic world.

Drawing very directly on Apichatpong's own memories of love and loss following the death of his father and the break-down of a relationship, and weaving together various forms of popular Thai storytelling traditions, *Tropical Malady*, like *Mysterious Object* and *Blissfully Yours*, foregrounds the memories and experiences of those on Thailand's social and political margins. Filmed on location in Petchaburi and Khao Yai national park, *Tropical Malady* is made up of two separate but interrelated stories. The first is the portrayal of a romance between Tong, a young male villager, and an army patrol soldier named Keng, set in a bustling small town where they go on dates to the movies, a restaurant, and the market, and spend time together in the countryside around Tong's family home. Arnika Fuhrmann argues that, in contrast to mainstream Thai cinema that tends to represent homosexuality as a form of 'damage' (both socially and individually), the film 'pursues the strategy of re-anchoring homosexuality in the mundane, public, and collective aspects of life in Thailand, in an affectively shaped social environment' (Furhmann 2008: 217). In my analysis of the film's sound design, I want to draw out this sense of affect to explore how this 'mundane, public, and collective' space depicted in the film can be understood as political. Foregrounding a sense of intimacy (both through the narrative and through the film's use of sound), the scenes I turn to now share a somatic and emotional appeal that transcends, or perhaps rather *circumvents*, language – privileging instead, the epistemology of embodied memory.

Tropical Malady begins with a black screen and a distinct hissing sound like static on old film – a juxtaposition that again heightens our awareness of the materiality of the film (made on 35 mm) and immediately foregrounds a sense of tactility. Reminiscent of the opening of a silent film, an inter-title reads:

"All of us by our nature are wild beasts. It is our duty as human beings to become like trainers who keep their animals in check and even teach them to perform tasks alien to their bestiality."

– Ton Nakajima

Suddenly, with an abrupt cut to a group of soldiers in bright daylight, we see that the men have found the body of a man in the long grass on the outskirts of the jungle and are posing with the corpse for macabre group photographs. Rather than using the convention of an establishing shot, beginning the film *in medias res* with an unstable, handheld camera momentarily *destabilises* the objectivity of the spectator's position by creating a rush of sensory stimulation. This effect is heightened by the constant loud 'whssshhhh' sound of the wind through the long grass and the men's bodies brushing against it – tactile, 'natural' sounds that form a stark contrast with the high-pitched, 'artificial' electronic beeping from their walkie-talkies and the digital camera against which they are juxtaposed. The soundscape tells us that this is no pastoral idyll.

The kinetic sounds of the men's bodies in this constantly moving environment are captured by the microphone (recorded on location at the time of filming) reaching our ears at what seems to be a slightly exaggerated level. The effect of this is to create a sense of immediacy or 'presentness' by heightening the phenomenological aspects of the scene. This is emphasised further by the close-perspective recording, which captures in rich detail the materiality of the diegetic sounds, such as those that convey the stiff, man-made texture and cumbersome weight of the tarpaulin (used to make a stretcher for the body) as the men drag it through the grass. The use of sound in this very *textured* way in this opening scene, by focussing on sound's materiality, establishes a pattern that will recur throughout the film in terms of the way it uses sound not simply to convey information, but to give the scene a sense of rhythm, and produce a feeling of intimacy between the film and the spectator. As the film cuts to an extreme long shot of the soldiers walking through the field, the sound of the wind continues to dominate the soundtrack while the image works in synchrony to capture the soft, undulating rhythms of the grass as the wind blows over it. Illustrating what Michel Chion terms 'visual microrhythms', which he describes as 'movements on the image's surface ... [which] create rapid and fluid rhythmic values, instilling a vibrating, trembling temporality in the image itself ... [affirming] ... a kind of time property to sound cinema as a recording of the microstructure of the present' (Chion 1994: 17), in this moment, narrative is momentarily suspended, as the aural and visual rhythms of the environment become the focus of attention. Through the synchronicity of sound and image, it is as though time itself has been paused to allow the spectator to contemplate the multiple, vibrating energies of the present moment creating a sense of Bergson's memory-time that draws together the different time frames of perception and recollection in a single instant.

A further layer to the 'texture' of the sound in this opening scene is added as a heavily distorted female voice transmitted through the walkie-talkie cuts through the wind: 'Forest fire on M. O. 12. Copy, over'. As one of the soldiers begins to flirt with her across the airwaves, their dialogue mimics a kind of hammy soap opera script, only eerily accented by sonic distortion and static electricity making their voices resonate uncannily around the field:

'Pretty Patcharee all alone. Do you need a friend?'
'I have lots of friends but my heart is still free.' (The soldiers respond humorously with smiles and camp 'oooohs!')
'Then I'll stop by. Don't be a stranger.'
'Is that Sawang? I've only heard your voice. I've never met you in person.'
'I'm at M.O.4. I hope my voice can soothe your heart. Can you sing us a song?'

As the soldiers walk away from a fixed camera towards the jungle, their voices begin to fade, becoming almost indistinguishable from the sound of the wind. Like the characters, the spectator struggles to catch the words:

'What? There's too much static – I can't hear you very well.'
'That's static from my heart. It's calling out to you ... Can I request a song?'
'Your battery might run out. I'll request a song from the radio. Is there a signal out there? ... This is for all you lonely guys. You're hot and wild like a forest fire.'

At this, the compressed sound of a hip-hop drum-fill (the intro of the film's theme song, 'Straight', by Thai band *Fashion Show*) is heard on the soldiers' radio. The low-angled camera dollies forward through the grass (like a predatory animal silently lurching forward to stalk its prey) and the sound becomes louder, gaining resonance as it shifts from the rather tinny diegetic sound emanating from the onscreen source of the radio to the more expansive, 'full' sound of the same song now heard on the non-diegetic soundtrack. When the soft, male vocals and lush guitar chords of the song's verse come in (echoing the sensuous sonic qualities of the wind through the grass), the image cuts to a long shot where the naked figure of a man crosses the frame from left to right, glancing briefly towards the direction of the now static camera. The figure then leaves the frame as the song's middle eight draws into focus an electronic arpeggio of melodic blips and bleeps that echo the soldiers' communication devices. As these sounds merge with the diegetic, ambient sounds of insect life, the screen fades up to a mid shot of Keng looking directly into the camera lens as he watches Tong's mother prepare a meal for her family and his patrol the night he and Tong first meet.

The sound design in the next part of the film imparts a building, pulse-like rhythm to a series of scenes shot in the style of observational documentary through the use of location sound recording. Importantly, in these scenes the materiality or 'concreteness' of the setting's quotidian soundscape is foregrounded and environmental sounds are given equal weight to dialogue. As Benedict Anderson has noted, in the first part of the film,

> there is no background music at all: instead the sounds of everyday country life, motorbikes, dogs barking, small machines working, and so on. The mostly banal conversations are also essentially 'background', and one does not need to pay careful attention to their content. Foregrounded are faces, expressions, body-language, silent communication with eyes and smiling lips. The elderly woman whom Tong calls Mae [mum] shows by her expression that she understands the courtship going on, but she says nothing about it, nor does anyone else in the village.
>
> (Anderson 2009: 161)

The soundscape moves from the hum of motorbikes and beeping car horns in the bustling street scenes, to the industrial sounds of percussive hissing and chipping, and the frenzied mechanical whirring of the machinery at the ice factory where Tong works. It then moves to a basketball game where he relaxes with his work-mates, to another noisy street scene, and then back to the factory. Orchestrated by Akritchalerm, this sonic sequence juxtaposes different elements of the ambient soundscape in order to build the rhythm and create an almost musical sense of phrasing that shapes both the mood and tone of the scene. This highly affective, multi-layered sound design, incorporating the extremes of environmental sound and street rhythm, articulates a sense of the developing relationship between the two men in place of dialogue.

These sounds and street rhythms reach a crescendo (literally, but also dramatically) in a scene that takes place in the rural outskirts of the town, after Keng has been trying to teach Tong to drive so he can get more work delivering ice. In a sudden heavy downpour, they run for cover underneath a sala (a wooden structure in the forest). Although the volume of the torrential rain on the soundtrack is extremely loud, at first their dialogue is just audible: 'I'm soaked,' Tong says. 'Are you cold?' Keng asks him, before handing him a gift that he has brought him (a copy of a cassette by

3. While the presence of
subtitles for non-Thai
speakers complicates this
argument, I would main-
tain that the general *feel*
of the scene in terms of
sensory impression
remains the same.

Thai pop band, *Clash*). In turn, Tong passes him a photograph from his wallet of himself when he was a soldier stationed at Kanchanaburi. Keng lies back and says, 'Ah, a soldier has a lonely heart'. Tong adds, 'You never die a natural death', to which Keng replies, 'I'd hate to die without having loved', gazing at the photograph. When Keng asks Tong who the other man is in the photograph, the camera draws back and films the scene in long shot. As the camera's position changes, so too does the point-of-audition which, in synchrony with the image, retreats to some distance away resulting in the dialogue being almost completely drowned out by the sound of the rain. The withdrawal of the microphone from the intimate space between them creates a narrative ellipsis – we do not find out who the figure was, and perhaps neither does Keng.

As I described at the beginning of this article, the vococentrism of classical cinema requires ambient sound to be lowered in the mix in order to ensure that dialogue remains audible. Although this is just one of many conceits used in film to maintain the illusion of naturalism, it is far removed from the reality of how the acoustic realm is perceived in lived space. By contrast, in this scene, as elsewhere in Apichatpong's work, 'natural' ambient sound appears heightened, to such a degree that the human voice has become inaudible. At this moment in *Tropical Malady*, while possibly quite authentic to the auditory experience of monsoons in Southeast Asia, to global cinema audiences (particularly in the North and West), the sound of the rain appears so exaggerated that it becomes almost *denaturalized*, thus heightening the sound's affective power through its disassociation from that which is signified: Schaeffer's 'sonorous forms' (Schaeffer 2005: 78). Rather than relying on language to convey information about the lovers' relationship, the sound design enables an embodied, phenomenological engagement with the sensuality of the scene, communicating 'feeling' trans-diegetically.[3] As the interplay of sound and image here produces a sense of synaesthesia (the sound and sight of raindrops running off the leaves might evoke the sense of touch for example), at the same time, it also performs a narrative function, expressing the unspoken erotic dimension to Keng and Tong's relationship.

The sense of temporal suspension witnessed at the beginning of *Tropical Malady*, in the field by the forest, occurs again later on, in a scene that signals the transition between the two parts of the film and similarly depends on affect to communicate aspects of the narrative. Standing in a deserted street at night after their date at the cinema (significantly, beneath Thailand's national flag and a yellow flag with royal or Buddhist insignia), Tong voraciously licks Keng's hand and fingers before disappearing into the darkness leaving Keng standing alone in stunned silence. On the soundtrack, the soft electronic arpeggios from the middle section of the film's theme song are heard again, this time acting as a bridge between the film's two halves as Keng drives through the night on his motorbike back to his base.

As mentioned earlier in relation to *Mobile Men*, long takes like this from the viewpoint of moving vehicles accompanied by joyous non-diegetic pop music are a recurrent visual and sonic motif in Apichatpong's work that convey a floating, dream-like state (Römers 2005). A similar sequence occurs in Apichaptong's previous film *Blissfully Yours*, a film that also follows a bifurcated structure, divided between urban and rural locations. The earlier film dramatises the experiences of Min, an illegal Burmese immigrant living in the border town of Khon Kaen in northeast Thailand. The first forty-five minutes are set in the stifling atmosphere of the town and feature a series of frustrating encounters in various bureaucratic settings. The second (following a belated credit sequence) takes place in the jungle of the Khao Yai National Park on the Thai-Burmese border. May Adadol Ingawanij and Robert Lowell MacDonald note that the film's opening scenes, in which all three central protagonists lie or evade the truth, portray the town as 'a site of intense alienation' that drives them to make their escape to the jungle in search of a release (Ingawanij and MacDonald 2006: 51). The transition between the urban and the rural takes place during an extended driving sequence through a winding country road and is accompanied on the soundtrack by a soaring, euphoric pop song by Thai singer Nadia that dominates the diegetic sounds of the car's engine and the surrounding environment. The music seems to denote the characters' transition into a state of bliss as they abandon the worries of their everyday lives.

In each of these transitional sequences, the 'dream-state' effect is underscored by the use of sound. Chion explains that 'suspension occurs when a sound naturally expected from a situation ... becomes suppressed, either insidiously or suddenly. This creates an impression of emptiness or mystery, most often without the spectator knowing it; the spectator feels its effect but does not consciously pinpoint its origin' (Chion 1994: 132).

Although noticeably out-of-kilter with the rest of the film, the effect is quite different when this void is filled by non-diegetic sound, as it is in these scenes in *Tropical Malady* and *Blissfully Yours*. Carole Piechota has discussed similar occasions in recent American films where a pop song dominates the soundtrack for the duration of the sequence. In these scenes, she contends, the affective power of the music is such that the cinematography becomes subordinate to the rhythm of the song, which momentarily controls the film's temporal register. In these moments, dramatic time is suspended, creating what she calls an 'audiovisual passage'. According to Piechota, these 'passages' elicit an affective response from the spectator where the music is not simply an expression of the protagonists mood or character, but rather seems to *transcend* the diegetic framework altogether, moving the spectator accordingly. As she argues, 'As these passages frequently last for several minutes (often the length of a pop song) and either lack or downplay dialogue, the perceiver is left with more time to acknowledge or contemplate her bodily and affective experiences' (Piechota 2009 cited in Shaviro 2010: 85).

As he rides through the town, for a minute and a half, the non-diegetic pop song that plays over this scene seems to simulate Keng's state of bliss and suggests that he is oblivious to events going on around him. This oblivion is registered sonically as the music obliterates all diegetic sound. Even as he passes the market place (which we remember from his date with Tong earlier as a vibrant, bustling and noisy place), the diegetic sound is completely, and surreally, absent. The song also effaces a temporal elision, providing a non-narrative bridge to the following morning when we see shots of Keng and his troupe being driven to their new placement. This transitional use of sound provides psychological insight as it suggests that Keng is still exhilarated from the previous night's erotic strangeness. Crucially, although this sequence is concerned primarily with feeling and pleasure, it also provides an example of what David Teh has termed the 'camouflage politics' of Apichatpong's work, as it (seemingly almost incidentally) draws attention to the social dissonance and violence that this bliss-state might conceal (Teh 2010). This is seen when Keng passes the marketplace and rides past a group of men brutally kicking a man in the stomach as he lies on the floor. They then run after the bike (where the camera is positioned) hurling bricks at it, apparently unnoticed by Keng.

In the second part of the film, entitled 'The Spirit's Path', the narrative totally abandons the quotidian sociability of the first half as it takes on the form of folklore, drawing on the long traditions of oral storytelling that cross the borders of Thailand, Burma, Laos, and Cambodia. The human voice is almost completely absent in this part of the film, which mostly takes place at night in near total darkness. Describing the difference in tone between the two parts of *Tropical Malady*, Jihoon Kim writes, 'The transposition of everyday life into dreamlike, mystical, and infinite land-scapes – a dense forest in *Blissfully Yours* and an obscure jungle in *Tropical Malady* – is accentuated by elliptical editing, extremely long takes (with fixed camera or smooth tracking), and the deep ambience of diegetic sound' (Kim 2010: 128). In contrast to the domestic and urban sounds-capes of the first part of the film, the 'deep ambience' of the second is achieved through Akritchalerm's combination of field recordings of jungle sounds and electronic effects. The jungle is bursting with sounds of life: animal calls, birdsong and insects buzzing. From somewhere in the darkness, a guttural cry is followed by the regular, rhythmic song of crickets that rises suddenly in pitch as if in response, and merges with the sound of hoarse barking coming from deep within the thick undergrowth. The crickets continue, the varying pace of their rhythm building tension like orchestral stridulation in a horror film: 'a sharp, high, slightly uneven vibrating that both alarms and fascinates' that, like tremolo, 'concentrates our attention. . .making us sensitive to the smallest quivering on the screen' (Chion 1984: 21, 20).

That this rich, affective soundscape draws primarily on field recordings (gathered by Akritchalerm), like the recordings of the jungle in *Uncle Boonmee Who Can Recall His Past Lives,* demonstrates a strong commit-ment to acoustic authenticity. Resonating with the film's overarching thematic concern with extinction and preservation, the use of these sounds can perhaps be understood therefore as a kind of 'acoustic archive'. This 'archive' preserves not only the sounds themselves (under threat from deforestation), but also a sense of the long-standing relationship felt between rural northeastern Thais and their auditory environment – what acoustic anthropologist, Steven Feld, has termed 'acoustemology' – a way of 'knowing and being in the world' through sound (Feld 2003: 226). In Benedict Anderson's essay on the reception of *Tropical Malady* in Thailand, he recounts a conversation he had with a friend from the northeast, who explained to him the centrality of sound in the lives of those who live near the jungle where animism is widely practiced. He explains that in the animist tradition, as humans can be reincarnated into animals after their death, the voices of deceased loved ones can be heard in the calls of birds and animals. As he explains, 'an uncle who died recently can be recognized in an owl hooting at night. When they sleep, people's spirits leave the body, and bring back messages [from the dead], sometimes in dreams' (Anderson 2009: 164). Although seeming to incorporate fantastical elements such as spirit guides speaking through animals, these elements and, in particular, the film's 'animist' use of sound, form part of a 'realist' narrative therefore, which is grounded in the *regional* experience, and understanding, of the acoustic realm. Drawing on the acoustemologies of Thailand's rural northeast, then, the film's sound design, like that of *Uncle Boonmee*, is able to capture an 'indexical trace' of the spirit world (Ingawanij 2009: 100).

While *Tropical Malady* deals mainly with personal memory, Apichatpong's fifth feature film, *Syndromes and a Century* blurs the boundaries of personal and social memories by connecting his family's story with that of a wider socio-political framework. The film is centrally concerned with the preservation of memory following the death of his father, and is based on Apichatpong's recollections of the stories told to him by his parents about their time working as doctors in a hospital before they were married. Their interrelated memories form the film's two halves (firstly from the perspective of a female doctor based on Apichatpong's mother, named Toey, and secondly an army trained doctor based on his father, called Nohng). As such, each half of the film resonates with feint traces and uncanny reflections of its double, as Apichatpong's recollections of his parents' memories form an elliptical and enigmatic narrative involving flashbacks and circular repetitions, formal devices characteristic of memory's representation on film discussed earlier. As Kong Rithdee notes, 'the elusive nature of what is inherited and what is actually remembered constitutes the enigma of *Syndromes and a Century*, a film in which time is fragmented and memories compartmentalised' (Rithdee 2007).

Like Apichatpong's earlier films, the use of structural bifurcation in *Syndromes and a Century* brings about a series of 'doubles' that foreground a sense of the fluidity of memory. Jihoon Kim argues that, the parallels and repetitions that occur between the two halves of the film form a cyclical pattern reminiscent of the effect of video looping in multi-screen video art which 'organize[s] the spatial arrangement and distribution of various temporal modalities – simultaneity, ellipsis, comparison, leaps into the future, the disparity between past and present, contestations between different viewpoints on a single event, and so forth' (Kim 2010: 135). Both halves of the film take place in hospitals: firstly, a small country hospital in Khon Kaen; and secondly, a large modern hospital in Bangkok. Hospitals and medical centres are another recurrent trope in Apichatpong's work and connote both physical and psychological healing. The English title of the film, *Syndromes and a Century*, however, extends this theme to a broader socio-historical framework, as Apichatpong's intimate family memories are made to resonate with those of the Thai nation: 'Everyone is a relative', we are told. While the first part of the film is dialogue-led and characterised by natural light, lush green plant life, and the natural, environmental sounds of its rural setting, the second, by contrast, features mostly artificial, fluorescent lighting, significantly less dialogue and a highly affective electro-acoustic score by Koichi Shimizu and Akritchalerm.

In an interview with Kong Rithdee, Apichatpong explains that the sound design for *Syndromes and a Century* was planned at the time of filming and was intended to communicate with the spectator on a visceral level:

> While filming at the hospital, the sound of construction pounded in my heart. Ideas for sound design developed in this way. And during editing, there were sound effects that I wanted to experiment with . . . I wanted [the sound to] resonate in the heart. I didn't want it to sound like a score or to have a clear melody, but to blend into the film's ambience. For the viewer to be aware of sound design but not be overly conscious − I tried to tune it to the same frequency as the viewer, to their heartbeat or their blood pumping so the sound is naturally absorbed into their body.
>
> (Weerasethakul 2008)

The notion of synchronicity between the film's score and the body of the spectator that Apichatpong describes here brings into focus ideas about bodily connection and sensory exchange that have been implicit in my analysis of his work thus far. It infers a sense of mutuality (symbolically, a dialogic and discursive space) that draws on the epistemology of embodied memory, a shared sense of how it feels to remember that is grounded in the senses. The insistent materiality of the film score in *Syndromes and a Century* (reminiscent of *musique concrète*) connects with, and extends, several of the traits found in the earlier films' sound design with regard to the relationship between sound, perception and phenomenology. The sound design towards the end of *Syndromes and a Century* becomes even more abstract however as heightened diegetic sounds merge with the electro-acoustic score, drawing out a symbolic connection between 'broken' bodies and a 'broken' state - a connection that has particular significance when considered against the backdrop of the political instability and violence that ensued in Bangkok after the 2006 coup d'etat which unseated Prime Minister Thaksin Shinawatra (Pongsudhirak 2008).

This connection occurs during a long sequence that begins with Nohng meeting a colleague who shows him around the hospital. When they go down the stairs to the basement (a hidden, subterranean space), the doctor explains to Nohng, 'the basement is reserved for military patients, war veterans and their relatives'. 'Everyone is a relative', Nohng answers. 'I know, small country huh?' he laughs. The film cuts to a workshop in one of the basement rooms where prosthetic limbs are being made and tried on by amputee patients. The mood changes abruptly as a disk drill whirs maniacally against the oppressive electric hum of the fluorescent strip lights above. A low, non-diegetic electronic drone slowly builds momentum, deepening in resonance as if seeping beneath these diegetic sounds like blood. This malevolent mood is underscored by the sound of rhythmic banging (perhaps of metal on metal) as the camera tracks slowly along the hospital corridor. Composed of multiple layers of industrial and electronic sound, enhanced by the use of reverb and delay, the tone of the soundscape here is both extremely melancholy and, at times, menacing, with the addition of a rhythmic electronic pulse that mimics the sound of a racing heart. With the absence of dialogue in this sequence in *Syndromes and a Century*, the deep resonance of the score communicates *through our bodies* a pain occurring on a national level ('everyone is a relative') that is unspeakable within the Thai public domain.

In the final moments of the film however, the unnerving sounds of the basement fade as the melodic voice of a woman singing outside accompanies a visual cut to the lake by the hospital. Nearby some people are waltzing, oblivious to the horror we have just witnessed in the basement. As the camera scans the scene, we see a large crowd doing aerobics to an upbeat pop song (the sound of jouissance in Apichatpong's films) and some monks playing nearby with a toy UFO. That *Syndromes and a Century* ends in this way, with such a startling contrast between the joyousness 'above ground' and the horror of the basement, complicates any simplistic 'reading' of the film through its sound design. And yet, that this seemingly innocuous moment was one of four scenes in the film that Thailand's Board of Censors demanded were cut is perhaps evidence in itself of a darker element in Thailand's political landscape and the control it has (or attempts to have) over its film culture (Ingawanij 2008).[4]

In *Background Noise: Perspectives on Sound Art*, Brandon LaBelle contends that '[t]heories of listening are often based on the notion of diffused subjectivity: through listening, an individual is extended beyond the boundaries of singularity ... toward a broader space necessarily multiple' (LaBelle 2006: 245). Through my analysis of the haptic soundscapes of Apichatpong's films, I have shown how the relationship between sound and intersubjectivity that LaBelle describes, can be achieved in film through particular recording techniques and uses of rhythm in the films' scores or diegetic soundscapes, and I have argued that sound design can create a feeling of intimacy and closeness through its appeal to the spectator's embodied self. A theory of spectatorship based on empathy and closeness that also acknowledges alterity marks a significant move away from (ocularcentric) psychoanalytic approaches that have, until recently, dominated film studies (such as found in the work of Jean-Louis Baudry, Christian Metz and Laura Mulvey). Connecting point-of-view with psychoanalytic theory – specifically, Lacan's 'mirror phase' – not only do these approaches seem to focus almost exclusively on the image, they also describe spectatorial engagement largely in negative terms, arguing that

4. The director refused to cut the film and instead, with support from several other key figures in Thailand's independent film scene, organised the 'Free Thai Cinema' campaign with the aim of reducing the state's power to ban and cut films. See Apichatpong Weerasethakul, 'The Folly and Future of Thai Cinema under Military Dictatorship', Thai Film Foundation, 08.11.07 http://www.thaifilm.com/articleDetail_en.asp?id=106 [accessed 24.01.13]

this process creates illusions of empowerment (Baudry 1974–75; Metz 1975; Mulvey 1975). Neglecting sound, however, means that they fail to address the many different ways that audiences connect with, and respond to, audio-visual media. Shifting the focus from the image to sound – and in particular to the *interplay* of sound and image – significantly enriches our understanding of how films make meaning and allows us to consider spectatorship in much more positive, potentially progressive, terms. In contrast to earlier 'visually orientated' models of spectatorship then, I hope to have shown through my analysis of these films how an exploration of the intersubjective and affective properties of film sound opens up the possibility of an ethical spectatorship based on the shared experience of listening.

SOURCES

Anderson, Benedict (2009), 'The Strange Story of a Strange Beast: Receptions in Thailand of Apichatpong Weerasethakul's *Sat Pralat*' in James Quandt (ed.), *Apichatpong Weerasethakul*, Vienna: Austrian Film Museum, pp. 158–177.

Ashmore, Jonathan (2000), 'Hearing' in Patricia Kruth and Henry Stobart? (eds.), *Sound*, Cambridge: Cambridge University Press, pp. 65–88.

Assarat, Aditya (2007), *Wonderful Town*, film, Thailand: Pop Pictures.

Barker, Jennifer (2009), *The Tactile Eye: Touch and the Cinematic Experience*, CA: University of California Press.

Baudry, Jean-Louis (1974–75), 'Ideological Effects of the Basic Cinematic Apparatus', *Film Quarterly*, 27:2, pp. 39–47.

Benjamin, Walter (2007), 'A Berlin Chronicle' in Peter Demetz (ed.), *Reflections: Essays, Aphorism, Autobiographical Writings*, New York: Schocken Books, pp. 3–60.

Callard, Felicity and Constantina Papoulis (2010), 'Affect and Embodiment' in Susannah Radstone? (ed.), *Memory: Histories, Theories, Debates*, New York: Fordham University Press, pp. 246–262.

Chion, Michel (1994), *Audio-Vision: Sound on Screen*, translated by Claudia Gorbman, New York: Columbia University Press.

Chulphongsathorn, Graiwoot (2006), 'Monster!: I survive through other people's memories', *Criticine*, www.criticine.com/feature_article. php?id = 35, accessed 18.03.08.

Coulthard, Lisa (2012), 'Haptic Aurality: Listening to the Films of Michael Haneke', *Film-Philosophy*, 16:1, pp. 16–29.

Dyer, Richard (2010), *Nino Rota: Music, Film and Feeling*, London: BFI & Palgrave Macmillan.

Feld, Steven (2003), 'A Rainforest Acoustemology' in Michael Bull and Les Back (eds.), *The Auditory Culture Reader*, Oxford: Berg, pp. 223–239.

Furhmann, Arnika (2008), 'Ghostly Desires: Sexual Subjectivity in Thai Cinema and Politics After 1997', PhD dissertation, University of Chicago, Illinois.

Ingawanij, May Adadol (2008), 'Disreputable Behaviour: The Hidden Politics of the Thai Film Act', *Vertigo* 3.8, Summer. http://www.vertigomagazine.co.uk/showarticle.php?sel=bac&siz=1&id=927 [accessed 24.01.13]

Ingawanij, May Adadol and Richard, Lowell MacDonald (2006), 'Blissfully whose? Jungle pleasures, ultra-modernist cinema and the cosmopolitan Thai auteur', *New Cinemas: Journal of Contemporary Film*, 4:1, pp. 37–54.

Ingawanij, May Adadol (2008), 'Sounds from life and the redemption of experience in Apichatpong Weerasethakul's films', conference paper presented at the Screen Conference, 5 July, University of Glasgow.

Ingawanij, May Adadol (2009), 'Observing Life's Remains', 55 Internationale Kurzefilmtage Oberhausen, Festivalkatalog, pp. 99–101.

Kawase, Naomi (2008), *Nanayomachi*, film, Japan and Thailand: JDCT.

Kim, Jihoon (2010), 'Between Auditorium and Gallery: Perception in Apichatpong Weerasethakul's Films and Installations' in Rosalind Galt and Kar Schoonover (eds.), *Global Art Cinema: New Theories and Histories* Oxford: Oxford University Press, pp. 125–140.

Kongsakul, Sivaroj (2010), *Eternity*, film, Thailand: Pop Pictures.

LaBelle, Brandon (2006), *Background Noise: Perspectives on Sound Art*, New York and London: Continuum.

Marks, Laura U. (2000), *The Skin of the Film: Intercultural Cinema, Embodiment, and the Senses*, Durham and London: Duke University Press.

Metz, Christian (1975), 'The Imaginary Signifier', *Screen*, 16:2, Summer, pp. 14–76.

Mulvey, Laura (1975), 'Visual Pleasure and Narrative Cinema', *Screen*, 16:3, Autumn, pp. 6–18.

Ong, Walter J. (2002), *Orality and Literacy: The Technologizing of the Word*, London and New York: Routledge.

Piechota, Carole (2009), 'Touching Sounds: The Audiovisual Passage in Contemporary Cinema', prospectus for PhD Dissertation, Wayne State University, Michigan.

Pongsudhirak, Thitinan (2008), 'Thailand Since the Coup', *Journal of Democracy*, 19:4, October, pp. 140–153.

Ratanaruang, Pen-ek (2006), *Invisible Waves*, film, Thailand: Fortissimo Films.

Ratanaruang, Pen-ek (2007), *Ploy*, film, Thailand: Fortissimo Films.

Ratanaruang, Pen-ek (2009), *Nymph*, film, Thailand: Fortissimo Films.

Rithdee, Kong (2007), 'Cinema of Impermanence' *Criticine*, www.criticine.com/review_article.php?id = 24, accessed 10.3.08.

Römers, Holger (2005), 'Creating His Own Language: An Interview with Apichatpong Weerasethakul', *Cineaste*, 30:4, Fall, pp. 42–47.

Schaeffer, Pierre (2005), 'Acousmatics' in Christopher Cox and Daniel Warner (eds.), *Audio Culture: Readings in Modern Music*, New York and London: Continuum, pp. 76–81.

Shaviro, Steven (2010), *Post Cinematic Affect*, Ropley, UK: Zero Books.

Sobchack, Vivian (2004), *Carnal Thoughts: Embodiment and Moving Image Culture*, Berkeley and London: University of California Press.

Suwichakornpong, Anocha (2009), *Mundane History*, film, Thailand: Electric Eel Films.

Teh, David (2010), 'On Sovereign Framing: Parergon in Southeast Asian Film and Video', conference paper presented at the 6th Association of Southeast Asian Cinemas Conference, Ho Chi Minh City, Vietnam.

Teh, David (2011), 'Itinerant Cinema: The Social Surrealism of Apichatpong Weerasethakul', *Third Text*, 25:5, pp. 595–609.

Weerasethakul, Apichatpong (2000), *Mysterious Object at Noon*, film, Thailand: Kick the Machine.

Weerasethakul, Apichatpong (2002), *Blissfully Yours*, film, Thailand: Anna Sanders Films.

Weerasethakul, Apichatpong (2004), *Tropical Malady*, film, Thailand: Anna Sanders Films.

Weerasethakul, Apichatpong (2006), *Syndromes and A Century*, film, Thailand: Anna Sanders Films.

Weerasethakul, Apichatpong (2007), 'The Folly and Future of Thai Cinema under Military Dictatorship', Thai Film Foundation, http://www.thaifilm.com/articleDetail_en.asp?id=106 [accessed 24.01.13].

Weerasethakul, Apichatpong (2008a), interview with Kong Rithdee, *Syndromes and A Century*, DVD, BFI.

Weerasethakul, Apichatpong (2008b), *Mobile Men*, film, Thailand: Dorje Films.

Weerasethakul, Apichatpong (2009), interview with 'Art for the World', http://art-for-the-world.blogspot.com/2009/01/interview-with-apichatpong.html, accessed 7.7.10.

Weerasethakul, Apichatpong (2010), *Uncle Boonmee Who Can Recall His Past Lives*, film, Thailand: Kick the Machine, Illuminations Films and Anna Sanders Films.

Weerasethakul, Apichatpong (2012), *Cactus River*, film, Thailand: Kick the Machine.

Weerasethakul, Apichatpong (2012b), *Ashes*, film, Thailand: Mubi.

CONTRIBUTOR'S DETAILS

Dr Philippa Lovatt, Digital Design Studio, Glasgow School of Art

Philippa Lovatt is a lecturer in Sound Theory at Glasgow School of Art. She gained her PhD in 2011 from the University of Glasgow. Her thesis 'Cinema's Spectral Sounds: Memory, History and Politics' analysed the use of sound in a group of important films produced in cultures of censorship by Bahman Ghobadi (Iran), Jia Zhangke (China) and Apichatpong Weerasethakul (Thailand). In response to the silencing of unofficial memories in authoritarian regimes, the study explored the politics of voicing and listening, drawing attention to the counter-discursive strategies at work in the films' different uses of sound, and suggested ways that we might think of film as testimony. A monograph based on the thesis is currently in development. Philippa is also a freelance film programmer and recently curated a season of Independent Chinese films as part of Takeaway China, an annual festival of Chinese Film and Photography in Glasgow.

Contact: philippalovatt@gmail.com

STEPHEN DEUTSCH AND LARRY SIDER

'The Soundtrack': The School of Sound Summer Workshop at the ifs internationale filmschule köln

ABSTRACT

In August, 2012, the School of Sound, in collaboration with the ifs (internationale filmschule köln), *conducted a three week post-production workshop, the object of which was to develop collaborative working practices between picture editors, sounds designers and composers. This article describes the processes involved and the philosophy of such interaction.*

KEYWORDS

sound design
film composition
film editing
collaboration
post-production

INTENTION

The original intention of the workshop was to create an understanding of how a film's soundtrack is related to the picture edit, and both are tied to the script. As the workshop evolved, however, its objective was to help practitioners to develop more integrated processes of production, derived through close collaboration with other key players in the post-production

The New Soundtrack 3.1 (2013): 81–88
DOI: 10.3366/sound.2013.0037
© Edinburgh University Press
www.euppublishing.com/SOUND

process. Part of the process involved participants collaborating with others from different countries and cultures. Specifically, the intention was:

> to integrate/tie together, both practically and conceptually, sound, music and image with narrative;
>
> to demonstrate how each component affects the other in a variety of ways;
>
> to enable the students to experience the practical and emotional effects of such collaboration;
>
> to demonstrate to students how ideas beget ideas; how a story can be formed from myriad sound and image impressions;
>
> to sharpen the students' focus – to encourage them to look increasingly deeply into what at first might appear to be a disappointingly simple (even 'boring') film and begin to develop an understanding of a growing yet subtle range of nuance, rather than just dealing with the surface story;
>
> to discover an alternative to the 'professional', commercial approach to storytelling and filmmaking – that can be integrated into a more traditional way of working.

BACKGROUND

Current industrial practices tend to separate and atomise the processes of editing, sound design and music. Such departmentalism is based on historical precedents, financial considerations and technological limitations. However, recent developments in convergent technologies have made it possible for editors, composers and sound designers to work in more collaborative ways. Indeed, it is possible (if not still somewhat clumsy) for a film's entire post-production process to be done on one software package. In smaller scale productions this practice is now becoming more common. However, the mind-sets of editors, sound designers and composers, especially those working in mainstream television and film production, present barriers to the adoption of collaborative systems of post-production. More interestingly, many practitioners when questioned, see the benefits of such collaboration, but bemoan the fact that 'it would be difficult to implement' in their own workplace. The workshop was designed to offer a model for such collaboration which could, hopefully, percolate into the working pattern of the participants.

PARTICIPANTS

The participants in the workshop, both tutors and students. brought with them a varied range of skills and experiences.

Tutors: The tutors were chosen for their professional filmmaking/ television experience as well as their extensive teaching skills practiced at the highest levels of film and media education. Their teaching reflected very personal methods as demonstrated in both their practical techniques and highly individual ways of thinking. Equally, the rationale of the course embodied the types of teaching promoted by the ifs, the School of Sound and the educational institutions to which the tutors are affiliated. The tutors were:

Prof. Stephen Deutsch – composer and sound designer; lecturer and Editor of *The New Soundtrack* journal; created the Music Design and Sound Design courses at Bournemouth University;

Graham Hartstone – re-recording mixer with a 46-year career in the Sound Department at Pinewood Studios; until his retirement he was Head of Post-Production at Pinewood; his mixing credits include *Eyes Wide Shut*, *Aliens*, *Thelma and Louise*, *Blade Runner* and 8 James Bond films; he now tutors part-time at the UK's National Film and Television School (NFTS), as well as at other film schools across Europe;

Peter Howell – film composer, lecturer at the NFTS and member of the BBC's Radiophonic Workshop;

Sylvia Ingemarsdotter – film editor best known for her long-time collaboration with Ingmar Bergman (*Fanny and Alexander*, *Autumn Sonata*, *Saraband*);

Pat Jackson – sound editor and film editor, Associate Professor in Cinema at San Francisco State University; her sound work ranges from Oscar-winning features (*Apocalypse Now*, *The English Patient*, *The Right Stuff*) to independent documentaries; she has enjoyed a long collaboration with Walter Murch on the films of Francis Ford Coppola, Philip Kaufman and Anthony Minghella;

Emil Klotzsch – studied at the *ifs* and now works as a sound designer, sound editor and mixer in Köln; Emil was the "hands on" sound mixer;

Annabelle Pangborn – film composer, sound designer and Head of Editing, Sound and Music at the National Film and Television School;

Larry Sider – Director of the School of Sound, Lecturer, Picture Editor and Sound Designer; former Head of Post-Production at the NFTS; co-editor of *The New Soundtrack* journal;

Diane Sider – Producer of the School of Sound; television producer and lecturer in business skills and disability equality.

Masterclass Speakers:

Birger Clausen – composer (*L'Italia Ci Appartiene*, *Lotus Eaters*, *The Last Thakur*);
Paul Davies – sound designer (*Morvern Caller*, *We Need To Talk About Kevin*, *Hunger*, *The American*);
Phil Parker – script writer and development consultant;
Monika Willi – editor (*The Piano Player*, *The White Ribbon*, *Amour*).

Providing a variety of points of view - rather than one way ("the" way) of solving each problem – was implicit in the teaching. In fact, more often than not the tutors disagreed amongst themselves on the ways to approach the film and its soundtrack. The aim of the teaching, however, was to inform the students' thinking, encouraging them to make final decisions within their creative team.

In addition to the day-to-day tutoring, four masterclasses gave wider perspectives on the roles of sound design, editing, composing and scriptwriting in the commercial industry. Phil Parker, (writer of the workshop film, *Empty Spaces*) began the workshop by explaining what he tries to do as a scriptwriter and how he sees the elements of a screenplay underpinning the decisions taken by the rest of the creative team from the directing through to post-production. The key being to understand the genre you are working in, and the intention of the writer in the choices they have made on the page. Birger Clausen, a young German composer and recent graduate of the NFTS, described how he has integrated the idealism of his film school teaching with the demands of commercial filmmaking. Sound designer Paul Davies explored how he had used his background in electro-acoustic music in creating tracks for such filmmakers as Steve McQueen, Lynne Ramsay and Anton Corbijn. Film Editor Monica Willi delivered the final masterclass, describing some of the methods she practiced to create rhythm and emotional colour by bringing sound into a picture cut, as evidenced in her work on the films of Michael Haneke. The three weeks were rounded off with sessions by Diane Sider, tutoring the students on how best to promote their technical and creative abilities to possible collaborators and employers.

Students: The fourteen students comprised young professionals and advanced students from nine countries. They had varying degrees of expertise in their disciplines, from basic operational skills to sophisticated and accomplished craft. Several of them had extensive professional credits. The variety of skills, outlooks and methods of working was both a benefit and a challenge to the running of the workshop. While the students clearly learned from one another, they needed to contend with creative partners who had specific ways of working based on their culture and professional training which varies from country to country.

Music: The four music students (from the UK, Italy, Germany and Azerbaijan) had extremely varied backgrounds and experiences. One was an accomplished professional conductor and composer in a mainstream idiom, another was accomplished in musique concréte, a third was well versed in modernist concert music techniques and the fourth was a performer/composer of mainstream vernacular music. Two had experience with computer sequencer equipment, one only with notational software and one had hardly any previous experience with computing. Each said that their motivation for coming onto the workshop was to gain the skills and insights which would enable them to make a career shift into film composition. Only one of them had ever composed to film, and that experience was transitory.

Editing: The four editors (from Germany, South Africa and Spain) were all experienced professionals. One had worked extensively in low-budget features in several countries and wanted to learn more about how sound affected both the pace and emotional impact of picture cutting. The second, a documentary editor, wanted to discover the new creative perspectives that could be found working openly with a creative team, without the restrictions of a commercial project. The third editor had worked primarily with shorts, art films and installations and was looking to increase his skills and outlook within a professional context. The fourth editor, from South Africa, wanted to develop a higher level of skill and creative thinking, to take part in a type of training which is not available in his country, and to share other people's experiences in a live, person-to-person context rather than through books or on the web.

Sound: The six sound designers (from the US, UK, Belgium, Russia and Italy) were of varied ages and experience. Two were advanced students with high levels of technical expertise: one studying film sound, the other doing a Research MA. Both were interested in the integration of sound and image in post-production from both conceptual and practical perspectives. The third was employed by a national television company, working on broadcast drama and documentaries. Another worked in a growing sound studio looking to acquire the skills to attract more high-level productions. The fifth was an award-winning radio producer and the last an aspiring sound designer with a background in electro-acoustic music who was looking to expand his understanding of film sound and establish contact with like-minded filmmakers.

FACILITIES

The workshop took place on the premises of the ifs, one of the leading European film schools, based in the centre of Köln. The ifs puts a special emphasis on the integration of sound with film editing and therefore was able to provide a well-equipped post-production environment. Each student had their own workstation which connected through the in-house network. Editors worked on either Avid or Final Cut Pro, the sound designers on ProTools (or had the option of editing on the same platform as their editor) and the composers on Logic. Voice, effects and Foley recording took place in a separate studio and the films were mixed in stereo on a ProTools system. All lectures and reviews were held in a central seminar room. The students had 24/7 access to the facilities with technical support throughout ordinary working hours.

THE ASSIGNMENT

Before arriving at the workshop students were sent the script to a film, specially written and filmed for the workshop (it had been used once before, in a similar workshop in Edinburgh in 2008). The film was written by Phil Parker and was entitled *Empty Spaces*. It was constructed in such a way as to offer the students (and viewers) an ambivalent story line. A man is killed in his flat, and his wife/partner is disappointed that the police cannot find the killer who, she is convinced, was having an affair with the victim.

The editors, sound designers and composers were tasked with creating a coherent narrative, a film of around 10 minutes' duration from the 100 minutes of rushes. There were to be four film versions to be delivered at the end of the three-week period, one each from teams consisting of an editor, a composer, and a sound-designer (two groups employed two sound designers working together).

Before their arrival at the workshop, composers were asked to provide music (temp tracks) to the script; their initial responses to the material. The music was laid onto the first cut of the film and reactions were given as to its suitability. Sometimes music which was inappropriate in one part of the film was completely effective somewhere else. For the composers, this was an especially interesting and unusual experience. They were required to take the music they had composed and reorganise it to fit the needs of the film – and these needs kept changing over the course of the editing process. Sometimes, it was necessary to abandon their first music entirely, and for the composer to develop a completely new take on the material. Similarly, sound designers began to assemble sound material and tentatively to lay it to picture (although this process began somewhat later in the gestation of the film).

THE SCHEDULE

The workshop took place over three weeks divided into

Week 1: picture edit, which included integration of rough sound and music cues
Week 2: develop sound design and score
Week 3: final mixes

The first day was devoted to introductory lectures including Phil Parker's which put the emphasis on narrative storytelling and how this works with post-production. Next came the rushes screening where the entire group viewed all the filmed material, a process that has largely died out. But the point was made that this was an important moment as it was the *only* time you see the rushes for the *first* time – like your audience. So first responses are crucial and should be noted.

In the following two weeks, most days began with a one-hour lecture by one of the tutors, followed by practical work and periodic reviews.

THE REVIEWS

At each stage of the process, reviews were held. These reviews discussed the work done so far, the efficacy of the outcome and the tasks which needed to be undertaken as a result. Some of the reviews were attended by a single team with tutors, others by everyone. For most of the students this was a completely new and daunting experience. In every case, we used two guiding questions for the students: 'What will the audience understand from this cut (or sound design, or musical gesture)?' and 'How does it make you feel?' This placed the emphasis not only on the creative intentions of the filmmaker, but more on the likely 'reading' of the film by the audience.

THE SOUND MIX

Graham Hartstone joined the group for the last week, where he attended all the final mix sessions. Students were encouraged to try both bold and subtle treatments of dialogue, music and sound effects' emphasis and balance in order to demonstrate various emotional influences on the storyline. It was stressed that the final soundtrack mixes must lead the first time viewer along the intended narrative, and support the picture edit with audio punctuations and reinforcements of the geography and timeline for each scene.

END PRODUCT

On the last day of the workshops each film was viewed, and the entire group was invited to discuss and analyse diverse aspects of each project. The four final films were all more than adequately completed. What everyone noticed was how far the films had evolved from the first cuts. In terms of emotion, the films gained a depth that the students had not anticipated. Most importantly, the development from early iterations to the lock-off were a combination of subtle, small changes in volume, placement and rhythm — not grand gestures. The difference made by a few frames or decibels became apparent. The answers to 'What does it mean?' and 'How does it make you feel?' came more easily and were more genuine than two weeks earlier. There was a sense of an alchemical process: something of value came from ordinary, base materials. Both students and tutors learned valuable lessons.

SUMMARY

By the end of the workshop there was a feeling of having experienced something that is hard to discover in the day-to-day workings of commercial film production. The students had glimpses of working methods that could help them create a different type of work, with more subtlety and nuance, that is achieved by close attention to each unfolding moment in a film.

As was the aim of this small, prescribed exercise, the relationship between a script's original intent and the final result could not only be seen but also felt, and this was largely due to the difference between 'being creative' and 'telling a story'.

However, the workshop also pointed out aspects of filmmaking that should be noted by educators and practitioners. Most obvious from the first day of the workshop was that filmmakers and film students have a surprisingly low tolerance for certain types of storytelling. For example, given the prevalence of the Hollywood model, students often seemed *not* to appreciate the benefit of staying with a shot or gesture, and the tendency to use rapid cutting was sometimes difficult to resist.[1] Similarly, especially for those editors who arrived with commercial experience, the impression is that the length of time that a student wants to work on a particular part of the editing process has become shorter, and that the patience needed to see, understand and 'connect' with material has been lost in the digital revolution and inertia of our daily lives.

1. Sylvia Ingemarsdotter's examples from Bergman's *Fanny & Alexandra* and especially *Autumn Sonata* went some way in influencing students about the power of the held shot and slow editing, allowing the audience time to 'decode' what they are seeing, and more importantly, feeling about what is being shown to them.

Initial student reactions to the film had been that the rushes were 'boring', 'ordinary' or 'unprofessional'. This judgement was based on the fact that the film contained little dialogue or obvious action. The rushes demand that the filmmakers construct their own versions of the story rather than merely follow the script in a shot-by-shot manner. Admittedly the film comprises, on the surface, a fairly simple story; the lighting and shooting is basic; the acting is more tv-ish than cinema. But the story and shoot have been designed in order to offer much leeway to the filmmakers. The complexity and emotional nuances of the story were there to be discovered and developed by the teams. Such discovery relies on a fair amount of experimentation by the creative teams. But as one sound designer said, 'I've never worked this way before!' So, for him, the work-shop was both exciting and annoying.

Current training and teaching of film encourages some collaboration and group-work but does not teach students what such collaboration actually entails (we took for granted that people will know how, and want, to collaborate). Collaboration is more than being in the same room together – it involves suppressing one's ego and an awareness of how one's own discipline and skills fit into the bigger picture. In the same way that some film composers realise that their music is not the main component of every scene, so must an editor come to terms with the fact that visuals can take a back seat to sound or music. The story can be told in many ways. One fault with our workshop is that we allowed the editors to take on their traditional role and start the process with a rough cut of the rushes to script. Immediately, picture had the upper hand. It took another $1\frac{1}{2}$ weeks for sound and music slowly to come to the fore.

Certain post-production processes, such as communal viewing of rushes or periodic reviews of work in progress, which professionals used to take for granted, were unfamiliar to some of the students as they are no longer performed in many film schools or even on professional produc-tions. And these are precisely the types of activities which promote reflection and collaboration.

There is also the 'problem' of having discovered a new way of working. For teachers and students alike it is a two-edged sword. Yes, the students took part in a collaborative process seeing how sharing and continually trading ideas is a truly creative process. But where can the filmmakers practice their new found methods as, for many of them, their usual work practices and industries will not allow for this type of collaboration?

ifs internationale filmschule köln – www.filmschule.de
The School of Sound – www.schoolofsound.co.uk